Computational Biology: An Introductory Text

Computational Biology: An Introductory Text

Ashley Scott

MURPHY & MOORE
www.murphy-moorepublishing.com

Computational Biology: An Introductory Text
Ashley Scott
ISBN: 978-1-63987-122-3 (Hardback)

MURPHY & MOORE
Published by Murphy & Moore Publishing,
1 Rockefeller Plaza,
New York City, NY 10020, USA

Cataloging-in-Publication Data

Computational biology : an introductory text / Ashley Scott.
p. cm.
Includes bibliographical references and index.
ISBN 978-1-63987-122-3
1. Computational biology. 2. Bioinformatics. 3. Systems biology. I. Scott, Ashley.
QH324.2 .C66 2022
570.285--dc23

For more information regarding Murphy & Moore Publishing and its products, please visit the publisher's website www.murphy-moorepublishing.com

Table of Contents

Preface

The branch of science which deals with the application and development of analytical and theoretical methods, computer simulation and mathematical modeling techniques is known as computational biology. It is involved in the study of ecological, social and biological systems. It is a multidisciplinary field which draws on the principles of applied mathematics, chemistry, biophysics, genetics, biochemistry and computer science. Computational biology makes use of biological data for the development of algorithms and models. These algorithms are used to develop an understanding of biological systems and their relationships. The various sub-fields of computational biology are computational biomodeling, computational anatomy, computational pharmacology, computational neuropsychiatry and computational genomics. Computational biology is an upcoming field of science that has undergone rapid development over the past few decades. Some of the diverse topics covered herein address the varied branches that fall under this category. This book, with its detailed analyses and data, will prove immensely beneficial to professionals and students involved in this area at various levels.

Given below is the chapter wise description of the book:

Chapter 1- Computational Biology is a branch of biology involving the combined application of math, statistics and computer science to study complex biological systems such as genetic sequences and the mechanism of protein unfolding. This chapter serves as an introduction and explains the subject matter in a simple manner. **Chapter 2**- Computational biology can be further divided into computational anatomy, computational genomics, computational neuroscience, computational pharmacology, computational neuropsychiatry, etc. These topics have been covered in a systematic and detailed manner in this chapter for the benefit of the reader. **Chapter 3**- Systems biology is an approach in biomedical research which uses analysis and modeling of complex biological systems. It employs a holistic approach and focuses on the smaller pieces. This chapter has been written carefully to provide an easy understanding of systems biology as well as its scope and impacts. **Chapter 4**- Gene expression is the process by which the genetic code of a gene is converted into a functional product, such as a protein for the production of cell structures. Gene regulatory network, biological network, transcription network, gene co-expression network, etc. are a few regulatory mechanisms through which gene expression is controlled. Gene expression and its various aspects are covered in detail in this chapter. **Chapter 5**- The mechanism used by cells to control the increase or decrease in the production of specific gene products is known as the regulation of gene expression. This allows the conservation of both energy and space with the cell. The various control mechanisms and genetic switches are closely examined and explained in an extensive manner in this chapter. **Chapter 6**- Bioinformatics refers to the interdisciplinary field that uses computation and analysis to understand and interpret biological data, which is an essential part of modern biology and medicine. Sequence analysis, sequence assembly, sequence alignment, BLAST, multiple sequence alignment, etc. fall under its domain. This chapter deals with bioinformatics and its related topics in detail.

At the end, I would like to thank all those who dedicated their time and efforts for the successful completion of this book. I also wish to convey my gratitude towards my friends and family who supported me at every step.

Ashley Scott

1

Computational Biology: An Introduction

Computational Biology is a branch of biology involving the combined application of math, statistics and computer science to study complex biological systems such as genetic sequences and the mechanism of protein unfolding. This chapter serves as an introduction and explains the subject matter in a simple manner.

Computational biology is the combined application of math, statistics and computer science to solve biology-based problems. Examples of biology problems are: genetics, evolution,cell biology, biochemistry.

Recent advancements in technology are enabling us to store an incredible amount of data. Initially, "Big Data" was perceived as a problem to be solved. In fact, we had reached a point in which we were able to store too much data without being able to make the best use of it. This drove a need for advancements in Data Science and Artificial Intelligence.

Nowadays, what was considered first as a problem has now become an open door to a world of innovations. Big Data has enabled many research fields such as Computer Vision and Deep Learning to flourish. This made possible for machines to perform complicated decision-making tasks and to extract from raw data information hidden, until then, to the human eye.

Biology is a subject which makes wide use of biological databases to try to tackle many different challenges such as understanding the treatment for diseases and cellular function. Datasets of biological data can be created from amino-acid sequences, nucleotides, macromolecular structures and so on.

Additionally, many robotic systems and algorithms frequently used in Computer Science are inspired by biological complexes. For example, Deep Learning Neural Networks are inspired in principle by the human brain structure.

Computational Biology Brings Order into Our Understanding of Life

"Biology adapted itself to the computer, not the computer to biology," writes Hallam Stevens in Life Out of Sequence, his ethnographic and historical account of computational biology. He explains: "Computers do not just scale up the old biology, they bring with them completely new tools and questions, like statistics, simulation, and data management that completely re-shaped the way biological research is being done."

One key example of how computers reshaped biological research is the use of databases and ontologies. Biological knowledge today is defined, organised, and accessed through computation. If Carl von Linné (also known as Carl Linnaeus), the Swedish botanist and father of taxonomy, lived today, he would be a computational biologist. As a botanist, he might take a leading role in a project like transplant to organise what we know about the genotypes and phenotypes of crops and model plants. Or he might work with the Gene Ontology Consortium to create shared vocabularies that unify biological knowledge across organisms. Just like Linné's Systema Naturae, such databases are key intellectual contributions to our understanding of life. Every other type of biological research builds on these foundations.

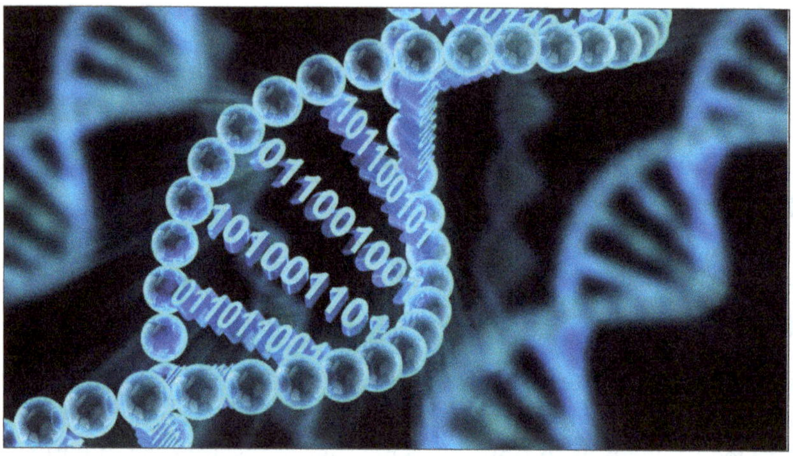

Another way computers have reshaped biology is by introducing statistics and data analysis methods. A good example understands how mutational processes shape genomes. Mutational processes—be it cigarette smoke, sunlight, or defects in homologous recombination—are not visible in individual mutations but only in their global patterns. How often is a C turned into a T? How does this frequency vary depending on the neighbours of the mutated base? How much of this frequency is explained by other features of the genome, like replication timing? Answering these questions helps us to understand basic properties of the mutational processes active in cells, and it is only possible by statistical techniques that identify patterns and correlations.

These types of analyses need large data collections, and thus the success of computational biology is closely linked to the success of large-scale efforts to gather genotypes and phenotypes of model organisms and humans. One of the first examples highlighting the power of computational approaches was sequencing the human genome, which showed how efficiently computational alignment and scaffolding methods were able to assemble the DNA fragments produced during shotgun sequencing, and modern Next Generation Sequencing techniques completely rely on advances in computational biology to analyse huge amounts of short sequence reads. DNA sequencing was once a Nobel Prize–worthy development. Now, computational biology is leading the way in turning it into a widely available and practical approach for both basic biology and medical research, which is currently revolutionising what we know about tissues and single cells.

Computational Biology Provides an Atlas of Life

By combining large data collections with databases and statistics, computational biology is providing a reference map for biology—an atlas of life that holds together individual insights. This map is not at the level of resolution provided by Google Street View, rather, it is a map like the one used by Columbus, Magellan, or Vasco da Gama—intrepid explorers in search of adventure. The map provides a general outline, but many areas are sketchy, and some important parts might even be missing and waiting for discovery. "Here be dragons," it just says. But even with all these shortcomings, the map is still an indispensable guide: the atlas of life provided by computational biology forms the background for planning, executing, and interpreting all focused small-scale experiments that probe the uncharted areas and push out the boundaries of biological knowledge.

Computational Biology Turns Ideas into Hypotheses

Finally, computers reshaped biology by making fuzzy concepts rigorous and testable. Here is one example: for decades, cancer researchers have discussed the idea that genetic heterogeneity between cells in the same tumour helps to make a cancer resistant to therapy. It is a simple idea: the more diverse the cell population is, the more likely it is that a subset of the cells is resistant to therapy and can regrow the tumour after all other cells were killed.

But how exactly can you measure "genetic heterogeneity," and how big is its influence on resistance development? To answer these questions, we had to turn the idea into a testable hypothesis. We can use genomic approaches to measure changes in cancer genomes at different sites in a patient and then defined quantitative measures of heterogeneity, which could be compared statistically to clinical information on treatment resistance. And indeed, we found evidence supporting the initial idea that heterogeneity determines resistance.

This is just one of many examples in which a quantitative computational approach was needed to turn a fuzzy idea into a testable hypothesis. Computational biology excels at distilling huge amounts of complex data into something testable in the wet lab, thus, shaping and directing experimental follow-up.

Computational Biology Algorithms

Some examples of algorithms used in computational biology are:

- Global Matching.

- Local Sequence Matching.
- Hidden Markov Models.
- Population genetics.
- Evolutionary Trees.
- Gene Regulation Networks.
- Chemical Equations.

Global Matching (also known as the Needleman-Wunsch problem) and Local Sequence Matching (also known as the Smith-Waterman problem) makes use of our knowledge about the proteins of an organism to understand more about other organisms proteins.

Markov Models are used for modelling sequences. In these types of models, the probability of an event to happen is just dependent on its previous state (this type of model can, for instance, be used to model a DNA sequence). Hidden Markov Models makes instead use of a probabilistic Finite-state machine in which, depending on the probability of the state we are in, we emit a letter and then move to the next state. The next state can possibly be equal to the original one.

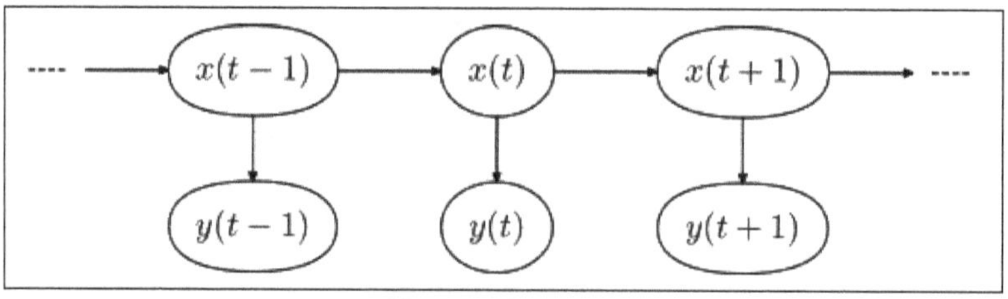

Hidden Markov Model.

Population genetics tries to model evolution. To do so, it commonly makes use of the Fisher-Wright Model. This model tries to simulate what happens at the location of a gene in selection, mutation and crossover conditions.

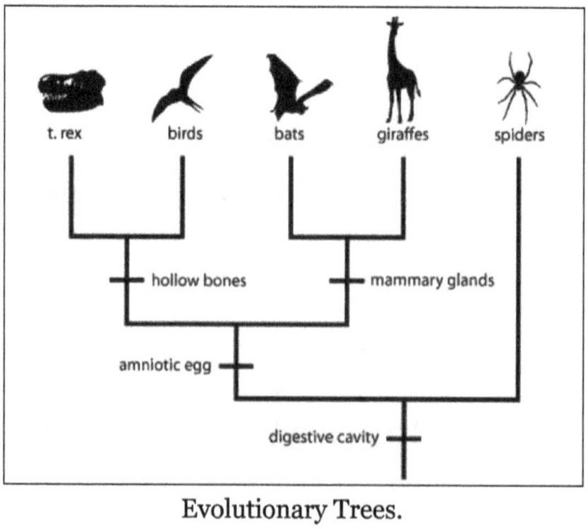

Evolutionary Trees.

Evolutionary trees can be created based on some form of evolutionary distance. There are two main types of evolutionary trees: distance-based trees and sequence-based trees. Evolutionary trees are used to explain distances between different species.

Gene regulation networks are formed thanks to the interaction of different proteins in an organism. The different proteins control's each other and according to the nature of their interactions, the cell type is determined.

Chemical equations can finally be used to describe the mechanics behind gene regulation networks. The reaction rates are dependent on the concentration of the elements in the chemical equations.

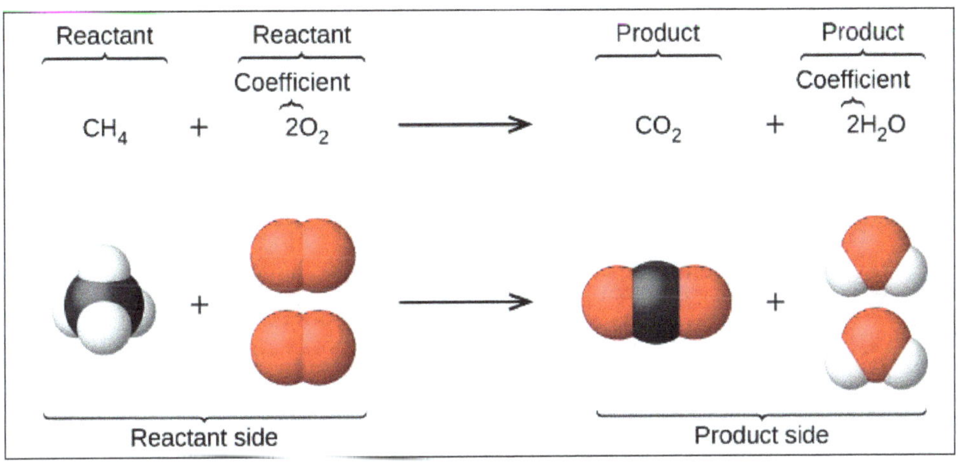

Chemical Equations.

Machine Learning for Biological Prediction

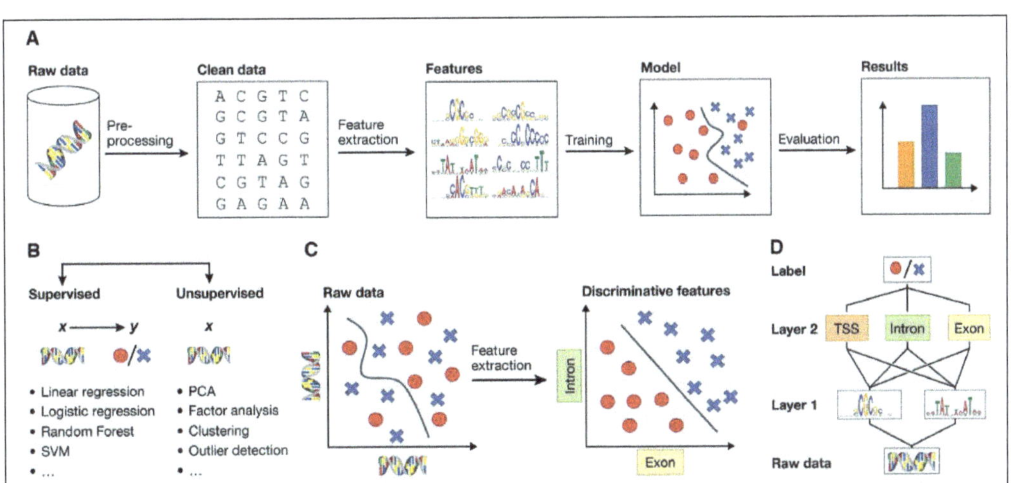

Machine Learning workflow in Biological Data Analysis.

Use of Machine Learning in Computational Biology is now becoming more and more important. Currently, applications are genomics (to study an organism's DNA sequence), proteomics (to better understand the structure and function of different proteins) and cancer detection.

Researchers demonstrated that using Convolutional Neural Networks (CNN) and Computer Vision, on cancer detection through image classification, can result in appreciable classification accuracy.

2

Sub-fields of Computational Biology

Computational biology can be further divided into computational anatomy, computational genomics, computational neuroscience, computational pharmacology, computational neuropsychiatry, etc. These topics have been covered in a systematic and detailed manner in this chapter for the benefit of the reader.

Computational Anatomy

Computational anatomy (CA) is rapidly emerging as a discipline focused on the precise study of the awesome biological variability of human anatomy. Although the study of structural variability of such manifolds can certainly be traced back to the beginnings of modern science, in his influential treatise bOn Growth and FormQ, Thompson had the clearest vision of what lay ahead, namely.

In a very large part of morphology, our essential task lies in the comparison of related forms rather than in the precise definition of each; and the deformation of a complicated figure may be a phenomenon easy of comprehension, though the figure itself may have to be left unanalyzed and undefined. This process of comparison, of recognizing in one form a definite permutation or deformation of another, apart altogether from a precise and adequate understanding of the original type or standard of comparison, lies within the immediate province of mathematics and finds its solution in the elementary use of a certain method of the mathematician. This method is the Method of Coordinates, on which is based the Theory of Transformations.

Thompson's vision is precisely the mathematical structure being studied in CA, a deformable template in which the space of anatomical imagery is an orbit under groups of transformations. In computer vision, the finite dimensional Lie group of rigid motions has been used for characterizing the orbits of object configurations. For the complexity of the biological shapes of human neuroanatomy, the infinite dimensional group of diffeomorphisms is used for characterizing the orbit of submanifolds of points, curves, surfaces, and subvolumes forming the anatomical configurations. The diffeomorphic transformations are constrained to be 1–1 and onto, and differentiable with differentiable inverse, so that connected sets remain connected, submanifolds such as surfaces are mapped as surfaces, and the global relationships between structures are maintained. For example, the brain is filled with evolutionarily stable gyral subsurfaces and subvolumes all of which have been categorized in the study of mammalian and human anatomy and are being quantified in CA.

Shown in figure is the methodology the CNA community is using for studying coordinatized manifolds on the brain via diffeomorphisms and bijections to local coordinates. The diffeomorphisms between R_3 provide smooth bijections to the submanifolds thereby allowing for the study of correspondences between different structures.

In CA many models are being examined by various investigators throughout the world, in which data enter through high-resolution dense imagery as well as sparse landmarked points, and shapes of manifolds directly. In almost all cases the study of anatomy is accomplished by identifying the anatomical picture with an unknown, hidden diffeomorphism on the background space of coordinates. There are essentially two models being examined in the CA community, the static metric mapping model and the dynamic metric growth model. For the static metric mapping model, the anatomical source is a deformable template $\mathcal{I}_\alpha$ corresponding to the orbit under the group $\mathcal{G}$ of one selected and fixed image $I_\alpha \in \mathcal{I}$. In the metric mapping model, a single observation from one time instant is observed, and the goal is to compare an exemplar to it via diffeomorphic transformation. In the second growth model, real-time evolution is observed of the coordinate system, and the evolution is to be inferred.

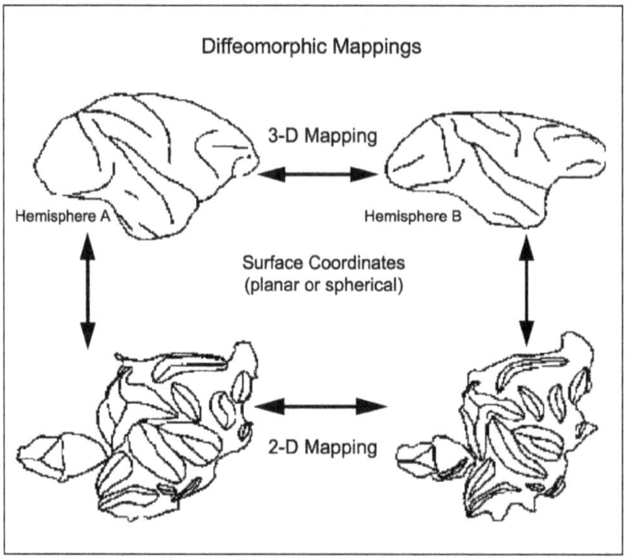

The top row shows the diffeomorphic mappings which are used for studying the three-dimensional of the brain, these mappings defined on sub-volume manifolds $X \subset R_3$. The bottom row shows diffeomorphic mapping between manifolds in $X \subset R_2$.

The greater CA community CA has three principal aspects: (i) automated construction of anatomical manifolds of points, curves, surfaces, and subvolumes; (ii) comparison of these manifolds; and (iii) the statistical codification of the variability of anatomical shape and structure via probability laws allowing for the inference of structural disease and functional responses on the geometries.

Towards (i) the construction of anatomical manifolds, the general area of the mathematical codification of biological and anatomical structure has expanded rapidly over the past several decades. With the automated construction of anatomical manifolds receiving tremendous focus by many of the groups represented in this special Neuroimage issue, atlases supporting neuromorphometric analyses are becoming available with many anatomical samples. Deformable and active models are being used for generating one-dimensional manifold curves in two to three dimensions. The differential and curvature characteristics of curves and surfaces have been examined as well with active and deformable surface models for the neocortex and cardiac systems. Local coordinatized representations for cortical manifolds have included both spherical and planar representations for local coordinates in studying the brain. A great deal of work has also been focussed on methods of segmentation of anatomical volumes into three-dimensional submanifolds. Automatic methods for maximum likelihood and Bayesian segmentation are being developed across the whole brain as well as focussing on particular gyri.

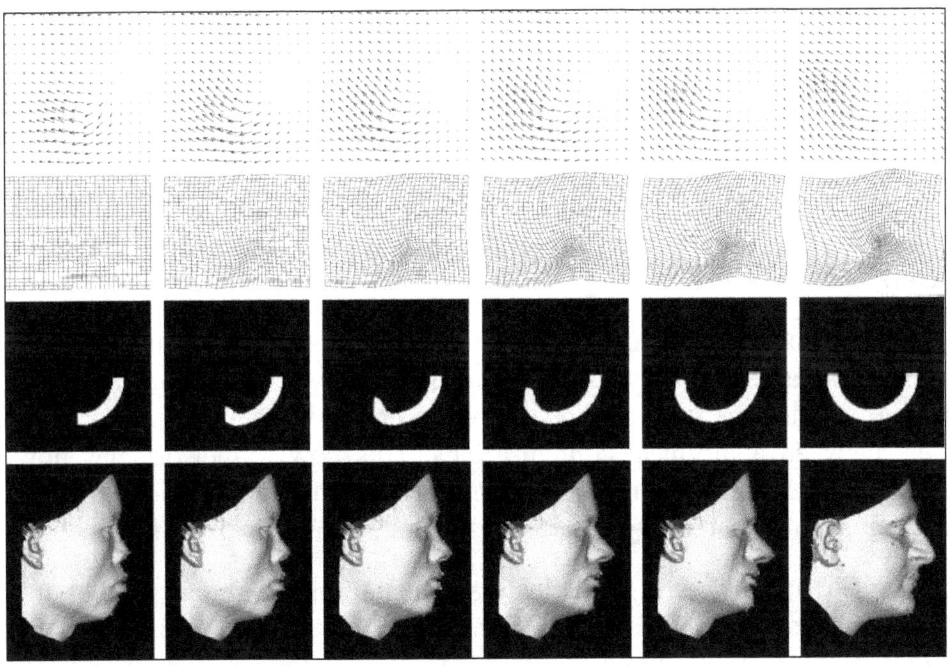

Figure shows (Row 1) Panels show the vector fields $v_t(\cdot)$, $t \in [0, 1]$ depicting the flow of a sequence of diffeomorphisms satisfying $dg_t/dt = v_t(g_t)$, $g_0 = id$. (Row 2) Panels depict the flow of diffeomorphisms g_t^{-1}, $t \in [0, 1]$ for the corresponding vector field times in the top row. (Row 3) Panels depict the flow of particles that carry black and white labels along the circular movement of the patch under the flow of diffeomorphisms $I \circ g_t^{-1}, t \in [0,1]$ for the corresponding times. (Row 4) Column 1 shows the starting template; column 6 shows the target surface that the diffeomorphisms were generated to flow to. Columns 2–5 show manifolds $M \circ g_t^{-1}$ along the geodesic generated by solving the diffeomorphic mapping of the faces.

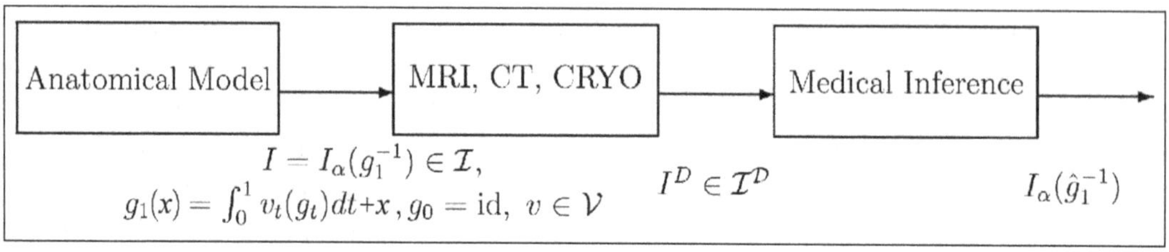

The source for the anatomical model is the set of all images $\mathcal{I}$ of the exemplars $I_\alpha(g_1^{-1})$. The observed data I^D are observations seen via the sensors (MRI, CT, cryosection), and are conditional Gaussian random fields with mean fields $I_\alpha(g_1^{-1})$. For the static metric mapping model, there is only one observation I^D the mean field of the exemplar $I_\alpha(g_1^{-1})$.

Towards (ii) comparison of anatomical manifolds, because of the fundamental structural variability of biological shape there has been great emphasis by groups on the study of anatomical shape via vector and metric comparison. The earliest vector mapping of biological coordinates via landmarks and dense imagery was pioneered in the early 1980s and continues today by Andresen et al., Bookstein, Avants and Gee, Bajcsy and Kovacic), Bajcsy et al., Dann et al., Gee, Gee and Bajcsy, and Gee and Haynor. Comparison via vector maps based on dense imagery is being carried on by many of the aforementioned groups. Most of the efforts define a preferred origin modeling deformations elastically based on small deformation vector maps on the dense volumes; vector mappings restricted to the cortical manifold are being computed as well.

Our own efforts in comparison of anatomical manifolds have been largely focussed on metric maps via large deformation diffeomorphic flows, originally proposed in Christensen et al.. Unlike the vector maps, these are not additive and provide guaranteed bijections between anatomical configurations. This flow framework has now since 1997 been developed into a complete metric space formulation, providing the mapping methodology with a leastenergy (shortest path) and symmetry property.

Concerning (iii) inference on statistical representations of shape, studies in CA of growth, atrophy, and disease have exploded over the past 5 years since the advent of automated methods for manifold construction. Applications to growth and statistical atlas building are evidenced by Andresen et al., Joshi et al., Paus et al., Sowell et al., Thompson et al., Toga et al., and Vaillant et al. Applications in CA to normal and abnormal aging in the dementia of the Alzheimer's type have been examined in both cortex and deep brain structures by many groups. As well, many investigators have been looking at the neuropsychiatric illness of schizophrenia. Researchers in CA continue to develop new methods of analysis in many of the neuropsychiatric illnesses including but not exclusive to ADHD, autism, major depressive disorder, psychosis, alcohol effects, neurologic disorders such as Huntington's disease, and multiple sclerosis, as well as comparison across illnesses.

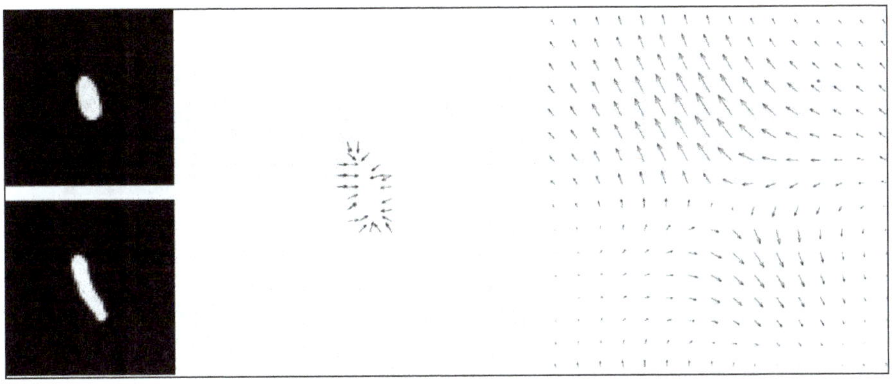

Figure shows Panel 1 shows the two binarized mitochondria sections I, I^D obtained from electronic microscopy. Panel 2 shows comparison between momentum at the identity $Av_0 = (I^D(g_1) - I)|Dg_1|\nabla I$ from Eq. ($Av_t = -\frac{1}{\sigma^2}\nabla\left(I \circ g_t^{-1}\right)Dg_{t,1}\left(I^D\left(g_{t,1}\right) - I\left(g_t^{-1}\right)\right), t \in [0,1]$) generated by the Beg LDDMM algorithm and the gradient of the image ∇I; arrows depict the direction vector of each. Notice how they superimpose almost exactly at every point; for the geodesic the momentum motion is normal as given by the conservation law. Panel 3 shows the vector fields v_0 at the identity given by the Green's kernel of the metric operating on the momentum Lv_0.

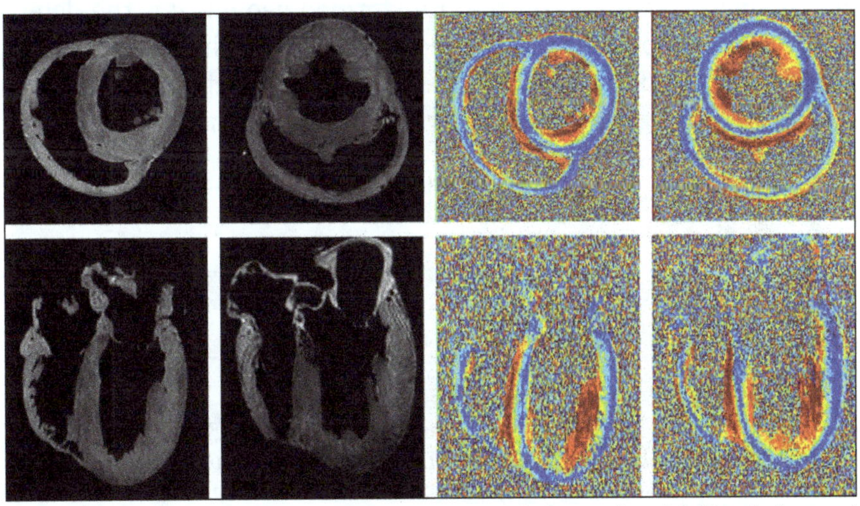

Columns 1 and 2 show hearts in grayscale MRI in normal (column 1) and failing (column 2) hearts; columns 3 and 4 show diffusion tensor images of those sections depicting orientation of the principal eigenvectors by color. The top row shows coronal sections, the bottom row shows axial sections.

The Anatomical Orbit Model: Diffeomorphisms Transforming Antomies

Now we examine the two basic mathematical structures that are used: the group of diffeomorphic transformations and the orbits of exemplars under diffeomorphic group action.

Diffeomorphisms Generated Via Flows

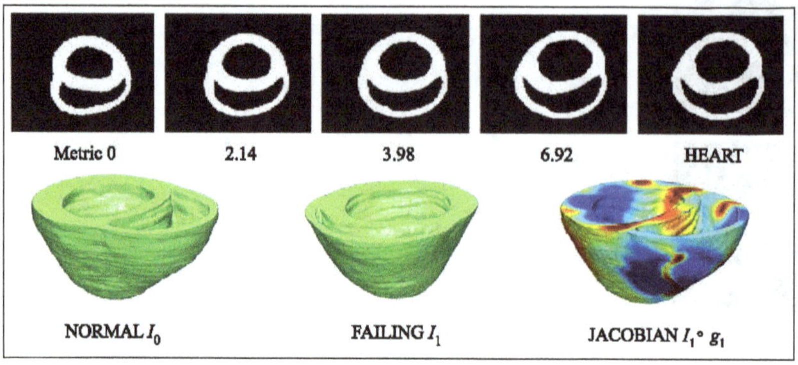

Figure shows Row 1 shows the sequence of metric maps $I(g_t^{-1})$ for t_0, t_6, t_{12}, t_{19}, target for two heart sections. Below each panel is the metric distance. Row 2 shows the geometries of the hearts in iso-surface-rendered images. Columns 1 and 2 show the normal template and target failing heart respectively. Column 3 shows the diseased heart mapped via LDDMM back to the template coordinates with the Jacobian superimposed. The heart data sets were first both rigidly registered and transformed via LDDMM based on input landmarks providing an initial condition for the diffeomorphism metric mapping algorithm. Right panel demonstrates the similarity of the transformed heart to the template geometry. Overlaid on the transformed geometry is the determinant of the Jacobian of the transformation at that point. Blue colors indicate regions where the determinant is less than unity indicating regions of contraction, whereas red regions are where determinant is greater than unity, indicating expansion.

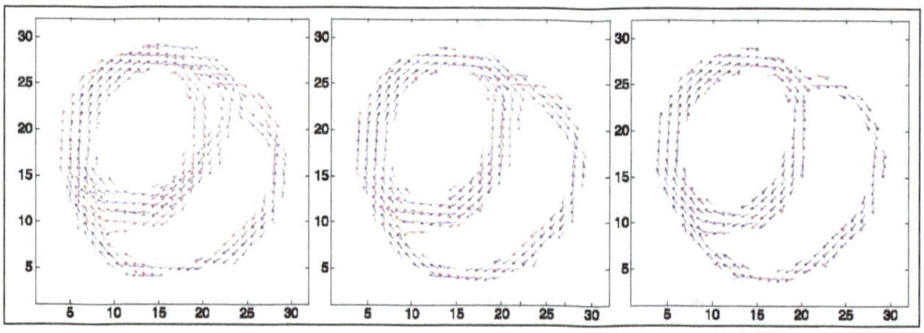

Panels show points along the geodesic for the LDDMM of the diffusion tensor vector data, with the red vectors showing the template and the blue vectors showing the target.

The most familiar diffeomorphisms are the matrix groups, which are finite dimensional transformations usually acting on points $x \in X = R^d$ identified with column vectors; the action is matrix multiplication on the vectors. Let X be the background space; the generalized linear group of scale, skew, rotation, and translation acts on the background space according to $(A, a)x = Ax + a \in X$, where A is a $R^d \times R^d$ matrix, $a \in R^d$; the law of composition is a semi-direct product $(A, a) \circ (B, b) = (AB, Ab + a)$. Computer vision has studied such matrix groups extensively for the representation of manmade objects. For biological shape, the action of the finite-dimensional matrix groups is extended to their natural analogue of the infinite dimensional diffeomorphisms g: X→X which are 1–1, and onto (bijection), continuously differentiable (smooth) with smooth inverse g^{-1}. The groups of transformations ($\mathcal{G}$; ∘) are a subgroup of diffeomorphisms generated via flows with law of composition of functions. Unlike the finite dimensional matrix groups, in the infinite dimensional setting, it is much less clear how to generate the mappings that are diffeomorphisms. For this, the approach is to construct the diffeomorphisms as a flow of ordinary differential equations (ODEs) as originally put forward for image matching by Christensen et al. Assume the diffeomorphisms g a G evolve in time as a flow g_t, $t \in [0, 1]$ with an associated vector field v: X x [0, 1] → R^d controlling their evolution, then the forward and inverse maps are given by,

$$\frac{\partial g_t}{\partial t}(x)=v_t(g_t(x)),\frac{\partial g_t^{-1}}{\partial t}(x)=-Dg_t^{-1}(x)v_t(x),x\in X,t\in[0,1]$$

With $g_0 = g_0^{-1}$ = id the identity map and D is the d d Jacobian matrix Df $\left(\frac{\partial}{\partial}\right)$.

Unlike for the finite-dimensional matrix group, the vector fields v_t, $t \in [0, 1]$ for the general diffeomorphisms are not a finite set of parameters and many things can go wrong. To ensure that the ODEs generate diffeomorphisms, the vector fields are constrained to be sufficiently smooth so that they integrate as a flow of diffeomorphisms. Dupuis et al. and Trouve´ have established that the technical condition on the minimal amount of smoothness required is that the fields be at least continuously one times differentiable $C_0^1(X,\mathbb{R}^d)$ with bounded sup-norm $\|v\|_{1,\infty} - \sup_{x\in X}|v(X)|+\sum_{i=1}^d\left|\frac{\partial v}{\partial x_i}(x)\right|$. This is accomplished by controlling the square integral of the derivatives so that the indexed family of vector fields v_t, $t \in [0, 1]$ are constrained to be in a Hilbert Sobolev space $V = H_0^s$ (zero boundary with s-derivatives having finite square integral); via the Sobolev-embedding theorems this then relates to the sup-norm pointwise continuity and differentiability. The construction of the diffeomorphisms group is as follows.

Construction: (Dupuis-Trouve´): Given vector fields v ∈ V on X with zero boundary, then if there exists some constant c, p ≥ 1, and for all $v\in V,\|v\|_V \geq c\|v\|_{p,\infty}$, with the vector fields satisfying $\int_0^1\|v_t\|_V^2\,dt<\infty$, then the ordinary differential equation Eq. ($\frac{\partial g_t}{\partial t}(x)=v_t(g_t(x)),\frac{\partial g_t^{-1}}{\partial t}(x)=-Dg_t^{-1}(x)v_t(x),x\in X,t\in[0,1]$) can be integrated over [0, 1], and $g_1(\cdot):X\rightarrow X$ is a diffeomorphism of X. The set of all such diffeomorphism solutions obtained from $\mathcal{V}=\left\{v:\int_0^1\|v_t\|_V^2\,dt<\infty\right\}$ forms the group:

$$\mathcal{G}(\mathcal{V})=\left\{g_t:\frac{\partial g_t}{\partial t}=v_t(g_t),t\in[0,1]:v\in\mathcal{V}\right\}$$

Clearly the identity diffeomorphism is in the group, inverse is in the group, and the composition of diffeomorphisms remains in the group.

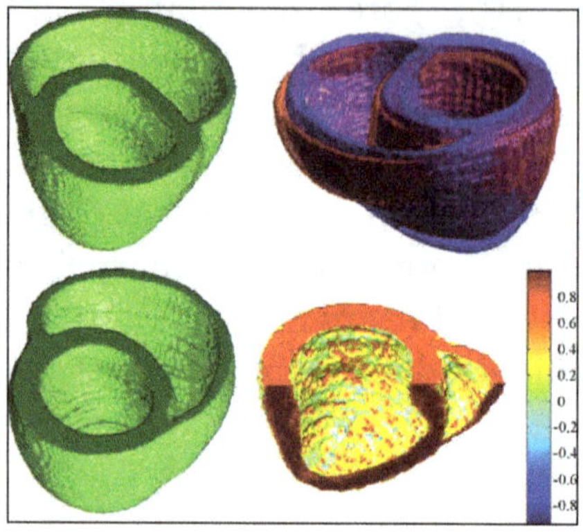

Column 1 shows the geometries of the two hearts. Column 2, top panel shows the two geometries superimposed after rigid motion; blue is template, red is target. Column 2, bottom panel shows the LDDMM solution color representing the dot product of the corresponding vectors after alignment. Red color means total alignment.

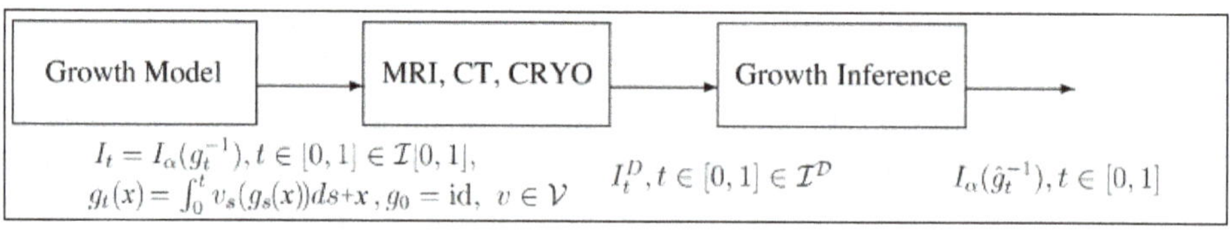

The source for the dynamic growth model is the set of all growth flows I[0, 1] of the exemplars $I_\alpha(g_t^{-1})$ $t \in [0, 1]$. The observed data I_t^D, $t \in [0, 1]$ are flows of observations seen via the sensors (MRI, CT, cryosection) and are conditional Gaussian random fields with mean fields $I_\alpha(g_t^{-1})$ $t \in [0, 1]$.

The Anatomical Orbit

In the deformable template model of computational anatomy, the images are the orbit under the diffeomorphic transformations G acting on exemplars or templates. Let the images be functions $I: X \to R^k$, scalar valued ($k = 1$ or vector $k > 1$) that associates to each point $x \in X$ a real-valued vector. The diffeomorphisms $\mathcal{G}$ act on the set $\mathcal{I}$ of all images via composition of the image with the inverse diffeomorphism.

Proposition: Defining for any $I \in \mathcal{I}$ and $g \in (\mathcal{G}, o)$, then $g.I = I \circ g^{-1}$ is a group action, with the orbit of imagery for the exemplar I_α given by $I_\alpha = \mathcal{G}{:}I = \{g.I_\alpha, g \circ G\}$.

Proof: That this is a group action follows from the fact that id.I = I and,

$$g(h,I) \doteq g.(I \circ h^{-1}) \quad (I \circ h^{-1}) \circ g^{-1} \quad I \circ (h^{-1} \circ g^{-1})$$
$$= I.(g \circ h)^{-1} = (g \circ h).I$$

The complete anatomy is the union over multiple orbits generated from different templates I_α, $\mathcal{I} = \cup_\alpha I_\alpha$. Throughout it is assumed that the exemplar or initial starting template is known and fixed. Figure illustrates the generation of a flow via the numerical integration of a specified series of vector fields $v_t(\cdot)$, $t \in [0, 1]$.

Diffeomorphisms Acting on Submanifolds

Almost all of the geometric work in CA concentrates on smooth submanifolds of R^d, as curved and surface submanifolds abound in human anatomy. Diffeomorphisms restricted to smooth submanifolds carry the submanifolds smoothly. If g is a diffeomorphism from $X \to X$ with $M \subset X$ a submanifold, then $g: M \subset X \to g(M) \subset X$ is a diffeomorphism from $M \to g(M)$. Shown in row 4 of figure are examples from Vaillant et al. demonstrating diffeomorphic mapping of surface submanifolds representing the human face.

The Metric Space of Diffeomorphisms and the Euler Equation on the Momentum Map

Although the matrix and infinite dimensional diffeomorphisms do not form a vector space and cannot be added, they form a group that is also a Riemannian manifold on which a metric space structure can be defined. The metric distance in the group of diffeomorphisms is the length of the shortest curved geodesic connecting them. This induces the distance between anatomical shapes induced by the distance between elements of the group that generate them. The Riemannian metric length is constructed defining curves g_t, $t \in [0, 1]$ on the manifold of diffeomorphisms connecting group elements $g_0 = g$, $g_1 = h \in \mathcal{G}(\mathcal{V})$, the curves are generated via the flow Eq. ($\frac{\partial g_t}{\partial t}(x)=v_t(g_t(x)), \frac{\partial g_t^{-1}}{\partial t}(x)=-Dg_t^{-1}(x)v_t(x), x\in X, t\in[0,1]$). The tangent element of the evolution $\frac{\partial g_t}{\partial t}(g_t^{-1})=v_t\in V, t\in[0,1]$ is a vector field, the tangent space a Hilbert space that is assumed throughout $(V,\|\cdot\|_V)$ to be a reproducing kernel Hilbert space with norm-square defined through the differential operator $A, \|f\|_V^2=\langle Af,f\rangle_2$,, with kernel K the essential inverse of A^1 satisfying,

$$\langle Kf,g\rangle v=\langle g,g\rangle_2$$

In the deformable setting it is natural to define the generalized momentum Av_t, $t \in [0, 1]$. The momentum is the function of the vector field that is integrated against the vector field to give the energy metric $\|v_t\|_V^2=\langle Av_t,v_t\rangle_2$. We assume throughout that v_t, $t \in [0, 1]$ is sufficiently smooth that Av_t is a function. The Euler equation describing the momentum evolution along the infimum length geodesic paths generalizes the finite matrix group Euler equation for shortest length paths.

Theorem: The geodesics connecting g, h $\in \mathcal{G}(\mathcal{V})$ minimizing $\int_0^1 \|v_t\|_V^2 dt = \int_0^1 \langle Av_t, v_t \rangle_2$ are solutions of the Euler equation,

$$\frac{\partial Av}{\partial t} + (Dv)^* Av + \text{div}(Av \otimes v) = 0$$

With $(\cdot)^*$ matrix transpose, divergence operator $\text{div}\, v = \sum_{i=1}^{d} \frac{\partial v_i}{\partial x_i}$ and div (Av ⊗ v) = (DAv)v + (div v)Av.

Proof of Theorem: The proof shown here parallels the finite Lie group approach shown to the author by Trouve´ and Younes. The Euler equation for infimum length paths generalizes the finite matrix group Euler equation. First, calculate the analogue of the Lie bracket for the finite-dimensional case describing how the vector field changes with a perturbation of the group flow as depicted in figure. Define g_t, $t \in [0, 1]$ to be the optimizing geodesic connecting $g_0 = g$ to $g_1 = h$; then each point along the curve is perturbed in the direction of η_t according to $g_t \to g_t(\varepsilon) = g_t + \varepsilon\eta t(g_t) + o(\varepsilon)$ where $o(\varepsilon)$ are terms vanishing at a rate faster than ε, with the perturbation satisfying the boundary condition $g_0(\varepsilon) = g$, $g_1(\varepsilon) = h$, implying $\eta_0 = \eta 1 = 0$, and $\eta(x) = 0,\ x \in \partial X$. The variational equation that the minimizing energy geodesic satisfies is,

$$0 = -\int_0^1 \|v_t + \varepsilon\partial_\eta v_t\|\ dt\Big|_{\varepsilon=0} = 2\int_0^1 \langle v_t, \partial_\eta v_t \rangle$$

To calculate $\partial_\eta v_t$ we have,

$$\frac{\partial}{\partial_t}(g_t + \varepsilon\eta_t(g_t) + o(\varepsilon)) = (v_t + \varepsilon\partial_\eta v_t)(g_t + \varepsilon\eta_t(g_t)) + o(\varepsilon)$$

$$\overset{(a)}{=} v_t(g_t) + \varepsilon Dv_t(g_t)\eta_t(g_t) + \varepsilon\partial_\eta v_t(g_t) + o(\varepsilon)$$

$$= \frac{\partial g_t}{\partial t} + \varepsilon\frac{\partial \eta_t}{\partial t}(g_t) + \varepsilon D\eta_t(g_t)\frac{\partial g_t}{\partial t}$$

$$\overset{(b)}{=} v_t(g_t) + \varepsilon\frac{\partial \eta_t}{\partial t}(g_t) + \varepsilon D\eta_t v_t(g_t)(g_t)$$

And equating (a) and (b), and applying g_t^{-1} on the right to both terms give,

$$\partial_\eta v_\eta = \frac{\partial \eta_t}{\partial t} + D\eta_t v_t - Dv_t \eta_t$$

Substituting Eq. ($\partial_\eta v_\eta = \frac{\partial \eta_t}{\partial t} + D\eta_t v_t - Dv_t \eta_t$) into Eq. ($0 = \frac{d}{d\varepsilon}\int_0^1 \|v_t + \varepsilon\partial_\eta v_t\|_V^2 dt\Big|_{\varepsilon=0} = 2\int_0^1 \langle v_t, \partial_\eta v_t \rangle$) gives,

$$0 = 2\int_0^1 \left\langle Av_t, \frac{\partial \eta_t}{\partial t} + D\eta_t v_t - Dv_t \eta_t \right\rangle_2 dt$$

$$\stackrel{(a)}{=} 2\int_0^1 \left\langle -\frac{\partial A v_t}{\partial t} - (Dv_t)^* A v_t, \eta_t \right\rangle_2 + 2\int_0^1 \langle A v_t, D\eta_t v_t \rangle_2 \, dt$$

With (a) following from integration by parts and the boundary condition $h_0 = h_1 = 0$. From Stokes' theorem, because v and η vanish on ∂X,

$$\langle A v_t, D\eta_t v_t \rangle_2 = -\langle \mathrm{div}(A v_t \otimes v_t), \eta_t \rangle_2$$

Giving,

$$0 = 2\int_0^1 \left\langle -\frac{\partial A v_t}{\partial t} - (Dv_t)^* A v_t - \mathrm{div}(A v_t \otimes v_t), \eta_t \right\rangle_2 dt$$

Because this is zero over all perturbations, it gives the Euler Eq. ($\frac{\partial Av}{\partial t} + (Dv)^* Av + \mathrm{div}(Av \otimes v) = 0$). Eq. ($\frac{\partial Av}{\partial t} + (Dv)^* Av + \mathrm{div}(Av \otimes v) = 0$) was first derived by Arnold for A = id for the divergence free setting in incompressible fluid mechanics; for our general setting, it was first proven using the calculus of variations.

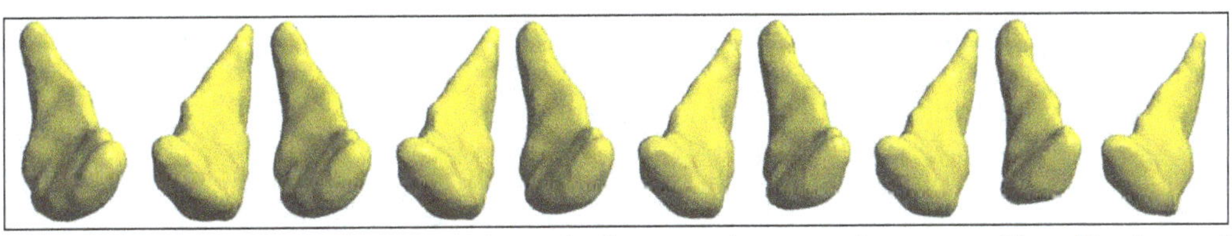

Series of volumes depicting aging in Huntington's disease patient over time as manifested in the caudate.

Large deformation diffeomorphic metric mapping In CA, observations are made on the underlying coordinates. Generally, the data are observed in multiple forms, including both measurements of material properties of tissue such as scalar and vector valued imagery derived from magnetic resonance imaging (MRI), CT, nuclear emission, and optical and acoustic scanners, as well as measurements made in the form of geometric properties of objects associated with submanifolds, curves, surfaces, and subvolumes. Throughout, assume the exemplar or initial starting template is known and fixed. The general model we pursue is depicted in figure.

Take the observable $I^D : X \subset R^d \to R^k$, in general a vectorvalued function. Define a distance measure between the underlying source or anatomical structure and the observable measurement $\delta : \mathcal{I} \times \mathcal{I}^D \to R^+$. We will generally study anatomy through the Gaussian random field model in which the observables $\mathcal{I}^D$ are modeled as a Gaussian random field with mean field $I_\alpha \circ g^{-1} \in \mathcal{I}$; the goal is to infer the diffeomorphism connecting the observation $\mathcal{I}^D$ to the exemplar $I_\alpha \in \mathcal{I}$ with associated matching term $\delta(I^D, I_\alpha \circ g^{-1}) = \frac{1}{2\sigma^2} \| I^D - I_\alpha \circ g^{-1} \|^2$ with σ^2 the weight corresponding to the noise in the measured Gaussian random field MRI image. As with many minimum norm variational problems, there are an infinite number of possible paths that minimize the distance function; we choose the ones that minimize the metric length and energy in the subgroup of diffeomorphisms solving the infimum,

$$\int_0^1 \|v_t\|_V^2 \, dt + \delta(I^D, I_\alpha \circ g^{-1})$$

As we shall now see, for metric mapping of images in which the data enter only through the two static observations at the start and endpoints of the flow, the geodesic is completely determined by the initial momentum. Along the geodesic, there is a conservation law, the momentum is constant. The particular form for the initial momentum Av_0 is determined by the matching problem associated with the template and target data as derived by Beg. This we term large deformation diffeomorphic metric mapping (LDDMM) of one shape onto another as originally solved. Now we calculate the momentum at the origin that solves the variational matching problem. Throughout, we understand the momentum in terms of weak equality, that is, through its action against test functions of smooth vector fields.

Theorem: Defining the vector field transported from the identity as $w_t = Dg_t(g_t^{-1})w_0(g_t^{-1})$, then along the geodesics satisfying the Euler Eq. ($\frac{\partial Av}{\partial t} + (Dv)^* Av + \operatorname{div}(Av \otimes v) = 0$) the momentum is conserved and determined by the momentum at the identity,

$$\frac{\partial}{\partial t}\langle Av_t, w_t\rangle_2 = 0 \text{ with } Av_t = (Dg_t^{-1})^* Av_0(g_t^{-1})|Dg_t^{-1}|, t \in [0,1]$$

[Beg's Large Deformation Diffeomorphic Metric Mapping] Given template I continuously differentiable with $\nabla(I \circ g) = (Dg)^* \nabla I(g)$, and $g_{t,u} = g_u \circ g_t^{-1}$, the momentum minimizing $\int_0^1 \|v_t\|_V^2 \, dt + \frac{1}{2\sigma^2}\|I^D - I(g_t^{-1})\|_2^2$ satisfying Eq.(previous) has momentum along the geodesic,

$$Av_t = -\frac{1}{2\sigma^2}\nabla(I \circ g_t^{-1}) Dg_{t,1}(I^D(g_{t,1}) - I(g_t^{-1})), t \in [0,1]$$

Proof of Theorem: Define the notation $w_t = (Dg_t w_0)(g_t^{-1})$ then,

$$\begin{aligned}\frac{\partial w_t}{\partial t} &= \frac{\partial}{\partial t}((Dg_t w_0)(g_t^{-1})) = \frac{\partial (Dg_t w_0)}{\partial t}(g_t^{-1}) + D(Dg_t w_0)(g_t^{-1})\frac{\partial g_t^{-1}}{\partial t} \\ &= (D(v_t(g_t))w_0)(g_t^{-1}) + D(Dg_t w_0)(g_t^{-1})(-Dg_t^{-1} v_t) \\ &= (Dv_t(g_t)Dg_t w_0)(g_t^{-1}) - D(Dg_t w_0)(g_t^{-1})v_t \\ &= Dv_t w_t - Dw_t v_t\end{aligned}$$

This implies,

$$\frac{\partial}{\partial t}\langle Av_t, w_t\rangle_2 = \left\langle \frac{\partial}{\partial t}Av_t, w_t\right\rangle_2 + \left\langle Av_t, \frac{\partial}{\partial t}w_t\right\rangle_2$$
$$\stackrel{(a)}{=} \left\langle \frac{\partial}{\partial t}Av_t, w_t\right\rangle_2 + \langle Av_t, (Dv_t w_t - Dw_t v_t)\rangle_2$$
$$\stackrel{(b)}{=} \left\langle \frac{\partial}{\partial t}Av_t + (Dv_t)^* Av_t + \mathrm{div}(Av_t \otimes v_t), w_t\right\rangle_2 = 0$$

With (a) from Eq. ($\frac{\partial w_t}{\partial t} = Dv_t w_t - Dw_t v_t$) and (b) the derivation leading to Eq. ($0 = 2\int_0^1 \left\langle -\frac{\partial Av_t}{\partial t} - (Dv_t)^* Av_t - \mathrm{div}(Av_t \otimes v_t), \eta_t\right\rangle_2 dt$) that by integration by parts gives the Euler equation equaling 0. The second part of the proof follows because $w_t = Dg_t(g_t^{-1})w_0(g_t^{-1})$ so that $(Dg_t)^{-1} w_t(g_t) = w_0$ with conservation of momentum implying,

$$\langle Av_t, w_t\rangle_2 = \langle Av_0, w_0\rangle_2 = \langle Av_0, w_0\rangle_2 = \left\langle Av_0, (Dg_t)^{-1} w_t(g_t)\right\rangle_2$$
$$= \left\langle (Dg_t)^{-1} {}^* Av_0, w_t(g_t)\right\rangle_2$$
$$= \left\langle (Dg_t)^{-1} {}^* (g_t^{-1}) Av_0 (g_t^{-1}) Dg_t^{-1}, w_t\right\rangle_2$$

Which reduces to Eq. ($\frac{\partial}{\partial t}\langle Av_t, w_t\rangle_2 = 0$ with $Av_t = (Dg_t^{-1})^* Av_0(g_t^{-1})|Dg_t^{-1}|, t \in [0,1]$) using the inverse function theorem $Dg_t^{-1} = (Dg_t)^{-1}(g_t^{-1})$.

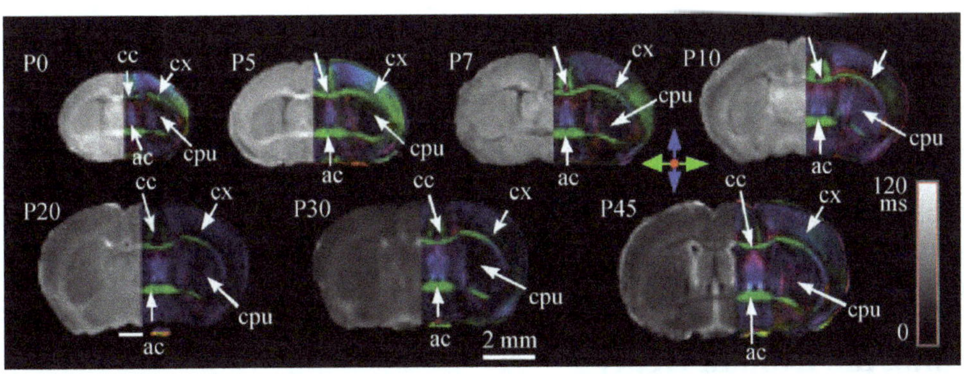

Figure shows MR images of postnatal mouse brains. Coronal MR T_2 and diffusion tensor images were shown at the level of anterior commissure. Images have been aligned to ensure proper orientation and position. Intensity in T_2 images is related to tissue properties, such as the content of myelin protein. MR diffusion tensor images reveal local tissue orientation. Both types of images were utilized in our mouse brain developmental study. The color scheme for diffusion tensor images was illustrated by color arrows in the figure, with red for local tissue whose orientation is anterior-to-posterior (perpendicular to the current plane), green for horizontal orientation, and blue for vertical orientation. The scalar bar represents 1 mm. Structure abbreviations are ac, anterior commissure; cc, corpus callosum; CPu, caudate putamen; CX, cortex.

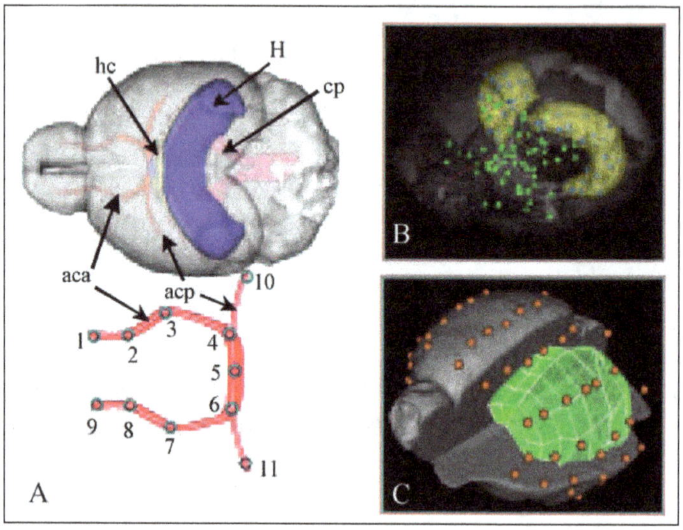

Figure shows (A) Landmarks defined on tubular white matter tracts. (B) Total of 60 landmarks defined throughout forebrain region (visualized as green dots); 44 landmarks defined inside the hippocampus (blue dots within yellow hippocampal surfaces in figure B). (C) Landmarks on cortical surfaces shown as orange dots in C. Structure abbreviations: aca, anterior part of anterior commissure; acp, posterior part of anterior commissure; cp, cerebral peduncle; H, hippocampus; hc, hippocampal commissure.

Beg's LDDMM algorithm discretizes Eq. $\left(Av_t = -\frac{1}{2\sigma^2}\nabla\left(I\circ g_t^{-1}\right)Dg_{t,1}\left(I^D\left(g_{t,1}\right) - I\left(g_t^{-1}\right)\right), t\in[0,1]\right)$ and solves it via relaxation. Notice at the identity $Av_0 = -\frac{1}{2\sigma^2}\left(I^D\left(g_1\right)-I\right)\left|Dg_1\right|\nabla I$. Shown in figure are comparisons between the momentum at the identity Av_0 and the gradient of the image ∇I. The momentum at the identity was calculated using the LDDMM algorithm to match via the geodesic flows and is compared to the momentum given by the normal to the level sets.

Mapping the Human Heart

Faisal Beg and Pat Helm working with Dr. Raimond Winslow have now mapped the three-dimensional geometries of normal and failing hearts whose geometry as well as fiber orientation is known. The results shown here are taken from Beg et al. Figure demonstrates anatomical variation including normal MRI sections (columns 1 and 2) and DTI section (columns 3 and 4).

Row 1 of Figure shows the sequence of metric maps $I\ (g_t^{-1})$ of the template to the target heart section demonstrating the use of the LDDMM algorithm. The operator in the metric was chosen to be $A = (-0.01\nabla^2 + id)^2$, and the flow $t \in [0, 1]$ was discretized into 20 time steps run with periodic boundary conditions. Plotted below the row of sections is the metric distance. Row 2 of figure shows the results of the three-dimensional LDDMM of the failing heart mapped to the template coordinates showing the geometries of the hearts in iso-surface-rendered images with superimposed Jacobian maps.

LDDMM of Diffusion Tensor Magnetic Resonance Heart Images

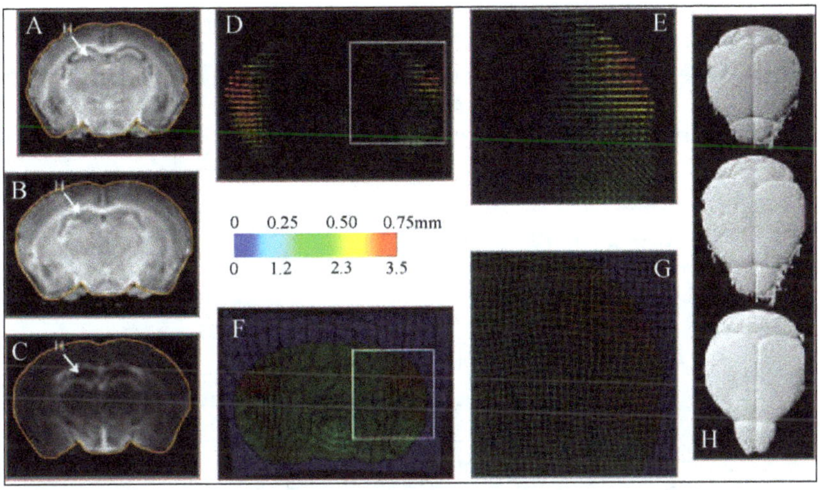

Figure shows LDDMM of P7 mouse brain images to P30 images. Coronal MR Images of P7 (A) and P30 (C) were shown with brain boundaries marked by orange curves. Deformed P7 (B) was overlaid with the brain boundary of P30. White arrows in A, B, and C are pointing to hippocampus (H). The transformation was visualized as color-coded vector plot (D) and Jacobian plot (F), with enlarged local areas shown in E and G, respectively. For vector plots, color of vectors represents the magnitude of local displacement. For Jacobian plot, color on deformed grids represents changes in local volume. Surface of P7, deformed P7 and P30 mouse brains were visualized in H, from top to bottom. Vector maps show how transformations deform a local voxel from its original location by taking the original location as the start of the vector and the destination the end; length and color of vectors reflect the magnitude of tissue displacement. The Jacobian map shows local volume changes; a value greater than unity corresponds to volume increase, a value less than unity correspond to volume loss, and unity for no volume change.

Diffusion tensor magnetic resonance imaging (DT-MRI) quantifies the anisotropic diffusion of water molecules in biological tissues, making it possible to non-invasively infer the architecture of the underlying structures. The measurement at each voxel in a DT-MRI image volume is a symmetric second order tensor. The principal direction of the diffusion tensor corresponds to the fiber orientation of the heart. We now examine the work of Cao et al. on the generation of LDDMM of the principal eigenvectors from DT-MRIs. Let the images be functions $I: X \to R^3$ that associate to each point a vector, the principal direction of the diffusion tensor (outside the object the vector is zero). Cao and Younes define the action $\mathcal{G}$ on the set $\mathcal{I}$ of all principal direction images as follows.

Theorem (Cao): Define for any $I \in \mathcal{I}$ and $g \in \mathcal{G}$,

$$g.I=\begin{cases} \dfrac{D_{g^{-1}}g\,I\circ g^{-1}\left\|I\circ g^{-1}\right\|}{\left\|D_{g^{-1}}g\,I\circ g^{-1}\right\|} & \text{when } I\circ g^{-1}\neq 0 \\ 0 & \text{when } I\circ g^{-1}=0 \end{cases}$$

With $D_f g$ denoting the Jacobian matrix of g evaluated at f. This is a group action.

Proof of Theorem: Clearly, id.I = I. For g and h in $\mathcal{G}$, $g.(h.I) = (g \circ h).I$ if $I \circ g^{-1} = 0$. If $I \circ g^{-1} \neq 0$,

$$g.(h.I)=\frac{D_{g^{-1}}g\left(h.I\circ g^{-1}\right)\left\|h.I\circ g^{-1}\right\|}{\left\|D_{g^{-1}}g\left(h.I\circ g^{-1}\right)\right\|}$$

$$=\frac{D_{g^{-1}}g\dfrac{D_{h^{-1}\circ g^{-1}}hI\circ h^{-1}\circ g^{-1}\left\|I\circ h^{-1}\circ g^{-1}\right\|}{\left\|D_{h^{-1}\circ g^{-1}}hI\circ h^{-1}\circ g^{-1}\right\|}\left\|I\circ h^{-1}\circ g^{-1}\right\|}{\left\|D_{g^{-1}}g\dfrac{D_{h^{-1}\circ g^{-1}}hI\circ h^{-1}\circ g^{-1}\left\|I\circ h^{-1}\circ g^{-1}\right\|}{\left\|D_{h^{-1}\circ g^{-1}}hI\circ h^{-1}\circ g^{-1}\right\|}\right\|}$$

$$=\frac{D_{g^{-1}}gD_{h^{-1}\circ g^{-1}}hI\circ h^{-1}\circ g^{-1}\left\|I\circ h^{-1}\circ g^{-1}\right\|}{\left\|D_{g^{-1}}gD_{h^{-1}\circ g^{-1}}hI\circ h^{-1}\circ g^{-1}\right\|}$$

$$=\frac{D_{h^{-1}\circ g^{-1}}(g\circ h)I\circ h^{-1}\circ g^{-1}\left\|I\circ h^{-1}\circ g^{-1}\right\|}{\left\|D_{h-1\circ g-1}(g\circ h)I\circ h^{-1}\circ g^{-1}\right\|}$$

$$=(g\circ h).I$$

Given two elements I and data I^D linked via $g_t = v_t(g_t)$, $t \in [0, 1]$, the LDDMM algorithm of Cao and Younes minimizes the matching function,

$$\int_0^1\left\|v_t\right\|_V^2 dt+\alpha\int\left\|I^D-g_1.I\right\|^2+\beta\left(\left\|g_1.I\right\|-\left\|I^D\right\|\right)^2$$

Where, α and β are constants. Cao et al. compute the optimizer performing steepest descent by computing the gradient with the step size e computed via golden section search for the line search. The vector fields generating the maps are integrated using second-order semi-Lagrangian schemes for the transport equation. Rigid motion was first calculated to provide global registration generated by aligning the centroids and the principal axes of the two objects. Figure shows sections along the geodesic as the algorithm aligns the vector images corresponding to the largest eigenvalue of the DT data. Shown in figure is the three-dimensional LDDMM-DT showing the geometries of the two hearts superimposed after rigid motion and LDDMM. The red color in panel 4 depicts almost total alignment as measured via the dot product of the corresponding principal eigenvectors.

LDDMM for growth and atrophy In the dynamic growth model, an observation time sequence is given and the goal is to infer the time flow of geometric change that carries the known exemplar through the growth period $t \in [0, 1]$. Here the orbits are flows of imagery, with growth corresponding to actual real-time, and comparison is performed by comparing the observables I_t, $t \in [0, 1]$ to its start point I_0. Here $I_0 = I_\alpha$ plays the role of the so-called template or exemplar. In static matching time is simulation time only, with the single target observable at time t = 1. Throughout we take as the dynamic growth model orbits of the exemplars under space-time flows,

$$\mathcal{I}_\alpha[0,1]=\left\{I_\alpha\left(g_t^{-1}\right),t\in[0,1]:\frac{\partial g_t}{\partial t}=v_t\left(g_t\right),g_0=id\right\}$$

Each element $I_\alpha\left(g\ \ \right),t\in[0,1]\in\ {}_\alpha[0,1]$ in the orbit is modeled as corresponding to a different growth trajectory of the exemplar. Figure depicts the basic static inexact matching model studied.

Preliminary observations suggest that the process of neurodegeneration in the disease is marked in the medial region of the caudate with considerable atrophy taking place in the head and the tail regions.

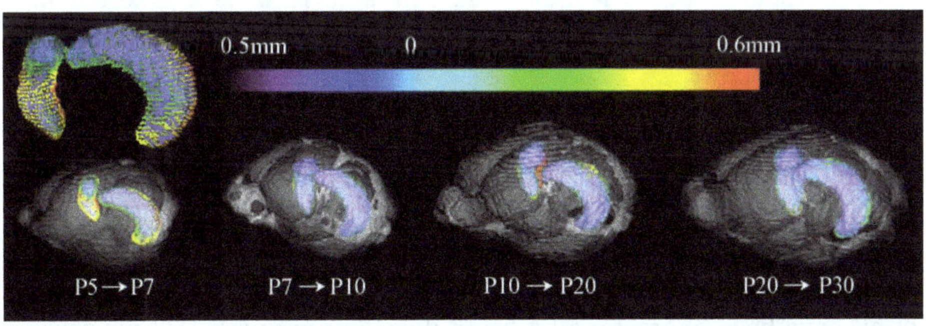

Outer-shape changes of mouse hippocampus measured using LDDMM. The displacement due to growth of hippocampus was color coded and the direction of growth was visualized by three-dimensional glyphs.

A central problem in CA is then to extract the hidden diffeomorphism sequence g_t, $t \in [0, 1]$ solving Eq. $\frac{\partial g_t}{\partial t}(x)=v_t\left(g_t(x)\right),\frac{\partial g_t^{-1}}{\partial t}(x)=-Dg_t^{-1}(x)v_t(x),x\in X,t\in[0,1]$ carrying the exemplar onto the static target observation sequence $I_t^D,t\in[0,1]$. The diffeomorphism is selected to minimize matching distance as a function of time $\delta=\frac{1}{2\sigma^2}\int_0^1\left\|I_t^D-I_\alpha\left(g_t^{-1}\right)\right\|_2^2 dt$.

We shall also examine the special class of sparse landmarked images for growth corresponding to finite dimensional landmarked shapes in which the images are a collection of points or landmarks, in I_N, the set of all collections of landmarks parameterizing the N-shapes IN $\in I_N$. Then the space of N-shapes $I_N = R^{dN}$ has metric $\rho : I_N \times I_N \rightarrow R^+$ defined by the lengths of the curves $g_t \in R^{dN}$; $t \in [0,1]$ bringing the points into correspondence. The matching distance function becomes $\delta = \frac{1}{2\sigma^2} \sum_{n=1}^{N} \int_0^1 \| y_{nt} - g_t(x_k) \|_{\mathbb{R}}^2 \, dt$.

Theorem: (Normal momentum motion for growth connection of dense imagery). Given the dense observable imagery growth sequence I^D_t, $t \in [0, 1]$ with the template I continuously differentiable with $\nabla (I \circ g) = (Dg)^* \nabla I(g)$ and $g_{t,u} = g_u \circ g_t^{-1}$, the momentum of the growth sequence minimizing $\int_0^1 \| v_t \|_V^2 \, dt + \frac{1}{2\sigma^2} \int_0^1 \| I_t^D - I(g_t^{-1}) \|_2^2 \, dt$ satisfies,

$$Av_t = -\frac{1}{2\sigma^2} \nabla \left(I \circ g_t^{-1} \right) \int_t^1 \left(I_u^D (g_{t,u}) - I(d_t^{-1}) \right) |Dg_{t,u}|$$

Given the sparse landmarks growth sequence y_{nt}, n = 1, 2,..., the momentum of the growth sequence minimizing $\int_0^1 \| v_t \|_V^2 \, dt + \frac{1}{\sigma^2} \sum_{n=1}^{N} \int_0^1 \| y_{nt} - g_t(x_n) \|_{\mathbb{R}}^2 \, dt$ is given by,

$$Av_t(x) = \frac{1}{\sigma^2} \sum_{n=1}^{N} \delta_{g_t(x_n)}(x) \int_t^1 \left(Dg_{t,u}(g_t(x_n)) \right) \times \left(y_{nu} - g_u(x_n) \right) du, x \in X$$

Where, δ is a Dirac evaluation functional.

The proof for the vector field can directly be found in Miller et al.; the momentum formulation is focused on in Miller et al. The algorithm is a gradient algorithm based on the discrete implementation of the equation. Notice for the static landmark case,

$$Av_t(x) = \frac{1}{2\sigma^2} \sum_{n=1}^{N} \delta_{g_t(x_n)}(x) \left(Dg_{t,1}(g_t(x_n)) \right)^* \left(y_n - g_1(x_n) \right), x \in X$$

Growth in Mouse

Zhang et al. have been characterizing brain development with DTI via three dimensional, digitized images of exvivo mouse brain samples with high resolution (100 Am per pixel) and high tissue contrast. These technical developments have paved the way for quantitative characterization of mouse brain anatomy during development. Brain development involves complex sequences of tissue growth and movements in three dimensions. During postnatal development, the mouse brain undergoes significant volume and shape changes. Figure shows example of the evolving mouse brain structures from the date of birth (P0) to postnatal day 45 (P45) in MR T_2 and diffusion tensor imagery. Volumetric measurements show that cortical and cerebellar volumes increased approximately 4- and 6-fold from P0 to P10, respectively. After P10, volumetric change stabilizes, while changes in shape can still be appreciated from P10 to P30. These results suggest that mouse brain development is a dynamic process with complex changes in both volume and shape being involved.

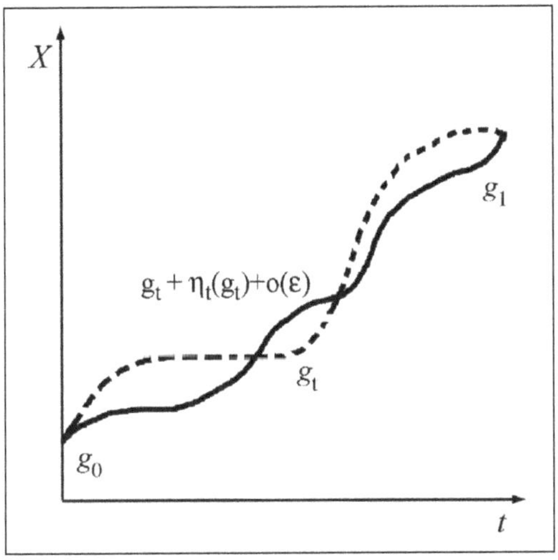

Panel shows perturbation of the flow $g_t \rightarrow g_t(\varepsilon) = g_t + \varepsilon h_t(g_t) + o(\varepsilon)$ in exact matching with $g_1(\varepsilon) = g_1$.

To track detailed morphological changes during development, we have been tracking cortical change via the introduction of many landmarks along structures of interest. As depicted in figure, landmarks are placed along several brain compartments including white matter tracts, hippocampus, cortex, and boundaries of white matter and cortex.

Figure demonstrates the LDDMM of a P7 mouse brain to a P30 mouse brain based on 270 landmarks that covered the entire forebrain region. The accuracy of the mappings was examined by plotting manually segmented tissue boundaries of P30 mouse brains over the deformed P7 mouse brain images. Fig shows the deformed P7 mouse brain images together with the outer boundary of the P30 mouse brain in orange. The P30 brain boundary agrees with the deformed P7 mouse brain images well. The shape of inner structures such as the hippocampus (H) also agrees with corresponding structures in the P30 brain images. The lateral cortical regions show significant growth from P7 to P30, with large tissue displacement coupled with near 3-fold volume increase. The average volume increase for the entire forebrain in this period is 2-fold.

Using LDDMM, various aspects of mouse brain growth may be characterized, including structural evolution and morphological variations. Figure shows the growth of the hippocampal surface measured using LDDMM methods. Displacements of the surface of the mouse hippocampus are computed and visualized. Although these results contain not only the effects of growth, but also the effects of landmark placement variations as well, it suggests that the hippocampal growth is not spatially uniform. The ventral part of the hippocampus experienced greater tissue displacement during development than the dorsal part.

Computational Genomics

Computational genomics is a subfield of computational biology that deals with the analysis of entire genome sequences. Transcending the boundaries of classical sequence analysis, computational genomics exploits the inherent properties of entire genomes by modelling them as systems.

Advances in the area of computational genomics produced substantial scientific results and established the field as an independent discipline. This challenges the utilitarian, `toolkit' attitude towards bioinformatics, and should result in a wider appreciation of this field in the experimental biology community. The wealth of approaches for the analysis of genomes is very extensive.

Most comparative analyses of biological systems today rely on extensive computational approaches such as the ones outlined below. We describe genome subtraction, whole-genome alignment, functional coupling and fusion analysis for the detection of protein interactions. All these approaches have the potential to increase our capacity to propose models of functional networks or associations beyond the threshold achieved by homology-based methods.

These methods depart from traditional computational methods described in the first part, because they primarily depend on the availability of entire genome sequences for the detection of patterns that rely on the organisation and Finiteness of genomes. Genome subtraction can only pick out unique genes if the genome sequence is completely known. Whole-genome alignment requires entire genomes, almost by definition. Finally, the precision performance of the last two methods crucially relies on completeness.

Genome Subtraction

Entire genomes allow the detection of `unique' sequences, genes that are not present anywhere in the database or in the close relatives of the species under investigation. These unique sequences may be the key determinants for species-specific phenotypic properties, such as pathogenicity, secondary metabolism properties and the like. These elements are sometimes components of cellular pathways that remain to be discovered and can be interesting drug targets in pathogenic organisms. To identify unique sequences, however, one has to detect equivalent (or `orthologous') genes, which are not always easy to define. Despite this shortcoming, this method will be most valuable for the comparison of bacterial strains or other, closely related species. Two interesting studies using this virtual subtraction method have appeared for Haemophilus influenzae and Helicobacter pylori.

Whole-Genome Alignment

Another area where technical advances resulted in some deeper understanding of the genome structure and thus function of certain species is whole-genome alignment. Previous systems could not cope with hundreds of kilobases of raw DNA sequence. This advance will facilitate detailed comparisons of genome organisation (revealing single nucleotide polymorphisms, translocations or inserts, repeats and syntenic regions in chromosomes). Another application is strain comparison, reminiscent of genome subtraction. This method has been applied to the comparison of two Mycobacterium tuberculosis strains, Mycoplasma genitalium and Mycoplasma pneumoniae and regions from mouse chromosome 6 against human chromosome 12.

Functional Coupling of Gene Clusters

Another method that exploits genome structure and organisation is the prediction of functional association of neighbouring genes. It has been observed that certain conserved gene clusters (which may be operons) contain functionally related genes. Thus, even for genes of unknown function, there is a possibility to predict their cellular roles or a more specific functional property on the basis of their neighbouring genes. Applications include the comparative analysis of two bacterial genomes, the comparison of nine bacterial and archaeal genomes to propose physical interactions of gene products and the use of gene clusters from 31 complete genomes to infer functional coupling and reconstruction of metabolic networks.

Fusion Analysis

Finally, based on the observation that the homologues of certain genes appear to fuse during the course of biological evolution, this approach attempts to predict functional association and protein interactions on the basis of gene fusion. The methods rely on the assumption that individual `component' proteins whose homologues are involved in a fused, multi-domain protein must be involved with each other in a protein complex, biochemical pathway or another cellular process. Detection of false positive predictions by this approach is difficult, mainly due to the lack of extensive experimental information about protein interactions.

Towards a Scientific Discipline

The explosion in computational analysis methods for complete genomes brought out not only technologies but also some key scientific results. For metabolic reconstruction, examples include the reconstruction of the metabolic networks of Methanococcus jannaschii, the analysis of the tricarboxylic citric acid cycle across a number of species, the characterisation of the known metabolic complement of E. coli, the distribution of functional classes across the domains of life and the prediction of functional networks in yeast.

For comparative genomics, examples include the detection of an archaeal genomic signature, the compilation of universal protein families, the comparison of three entire eukaryotic genomes, the detection of eukaryotic signalling domains in archaea and bacteria, the distribution of individual protein families across species, the patterns of protein fold usage in microbial genomes, the derivation of the universal tree based on enzyme families and the derivation of species relationships based on gene content.

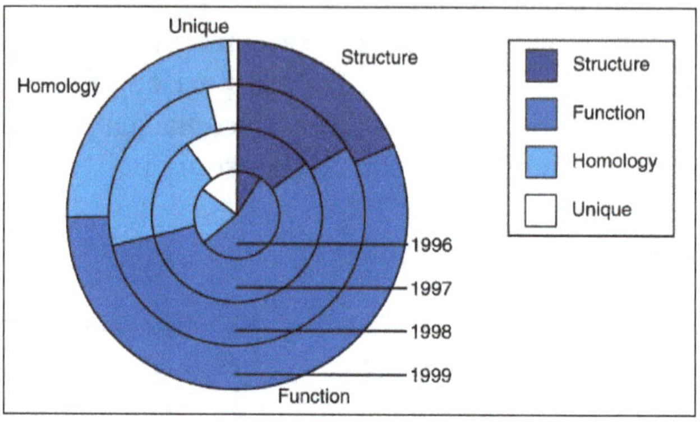

Figure shows the `function bottleneck'. Information clock illustrating the improvement of annotations identified for a given genome over the years 1996-1999. The four levels of annotation range from homologues of known structure (blue) and homologues of known function (marine) to homologues of unknown function (cyan) and unique sequences (white). Note that although structure and homology increased over the years, the function prediction level stalled.

Taking a closer look at the properties of entire genome sequences, two things become apparent: First, comparative analysis greatly enhances our abilities to `predict' and detect molecular function using sequence information and second, the current bottleneck in genomics appears to be the turnover of experimentally obtained novel properties for molecular families of unknown function. Once the `function universe' is covered, it may be that computation will acquire a truly central role in biological science.

Computational Neuroscience

Computational neuroscience is the field of study in which mathematical tools and theories are used to investigate brain function. It can also incorporate diverse approaches from electrical engineering, computer science and physics in order to understand how the nervous system processes information.

Computational Neuroscience works to identify dynamic neural networks to understand the principles that govern neural systems and brain activity potentially related to information processing and brain disease. The quantitative nature of the field is primarily concerned with complex computational analysis of electrical and chemical signals in the brain to understand the role of neurons in the processing of information using mathematical models at various spatial-temporal scales. These neural models provide experimental data that are used to generate new hypotheses that can further be tested by biological or psychological experiments.

Importance of Computational Neuroscience

Research studies in the field of computational neuroscience are complementary with psychology in understanding the nature of behaviors. Whereas a psychologist might attribute behavior to certain attitudes or emotions, a computational neuroscientist might interpret that behavior concerning the efficiency that certain groups of neurons propagate information in a specific context within the brain. They view the brain as a black box – there are numerous inputs and environmental variables introduced, something happens inside, then the behavior is performed. Computational neuroscience attempts to simulate brain function to find the unknown factors that influence human behavior.

As computational neuroscientists advance their understanding of the structure of the human brain at cellular and molecular levels, various fields of neuroscience develop and innovative applications begin to emerge. Important frontiers of computational neuroscience include the rapidly expanding field of artificial intelligence and machine learning, medical sciences, and human psychology. Further into the future, neuroscientists predict applications in optimizing computer chips, constructing very-large-scale integration (VLSI) architectures with neuromorphic engineering, advancing Human Machine Interface (HMI) technology for the handicapped and more.

Relationship between Computational Neuroscience and Machine Learning

Machine learning is the scientific study of statistical models and algorithms used to automate computer systems to perform specific tasks without explicit instruction. Some computational neuroscientists have attempted to integrate this technology into their research by developing new data analysis techniques that would allow them to analyze larger and more complex datasets that they otherwise wouldn't be able to interrogate. However, these two terms should not be used interchangeably. For example, many computational neuroscience experimentalists are focused on creating dynamic neural network models that are as biologically realistic as possible. This aspect of computational neuroscience does not align with machine learning, whereas optimizing mathematical insights within data regarding how the brain stores information might provide a productive overlap.

Research Topics of Current Computational Neuroscience

Single-Neuron Modeling

Even single neurons have complex biophysical characteristics and can perform computations (e.g.). Hodgkin and Huxley's original model only employed two voltage-sensitive currents (Voltage sensitive ion channels are glycoprotein molecules which extend through the lipid bilayer, allowing ions to traverse under certain conditions through the axolemma), the fast-acting sodium and the inward-rectifying potassium. Though successful in predicting the timing and qualitative features of the action potential, it nevertheless failed to predict a number of important features such as adaptation and shunting. Scientists now believe that there is a wide variety of voltage-sensitive currents, and the implications of the differing dynamics, modulations, and sensitivity of these currents is an important topic of computational neuroscience.

The computational functions of complex dendrites are also under intense investigation. There is a large body of literature regarding how different currents interact with geometric properties of neurons. Some models are also tracking biochemical pathways at very small scales such as spines or synaptic clefts.

There are many software packages, such as GENESIS and NEURON that allow rapid and systematic in silico modeling of realistic neurons. Blue Brain, a project founded by Henry Markram from the École Polytechnique Fédérale de Lausanne, aims to construct a biophysically detailed simulation of a cortical column on the Blue Gene supercomputer.

Modeling the richness of biophysical properties on the single-neuron scale can supply mechanisms that serve as the building blocks for network dynamics. However, detailed neuron descriptions are computationally expensive and this can handicap the pursuit of realistic network investigations, where many neurons need to be simulated. As a result, researchers that study large neural circuits typically represent each neuron and synapse with an artificially simple model, ignoring much of the biological detail. Hence there is a drive to produce simplified neuron models that can retain significant biological fidelity at a low computational overhead. Algorithms have been developed to produce faithful, faster running, simplified surrogate neuron models from computationally expensive, detailed neuron models.

Development, Axonal Patterning, and Guidance

Computational neuroscience aims to address a wide array of questions. How do axons and dendrites form during development? How do axons know where to target and how to reach these targets? How do neurons migrate to the proper position in the central and peripheral systems? How do synapses form? We know from molecular biology that distinct parts of the nervous system release distinct chemical cues, from growth factors to hormones that modulate and influence the growth and development of functional connections between neurons.

Theoretical investigations into the formation and patterning of synaptic connection and morphology are still nascent. One hypothesis that has recently garnered some attention is the minimal wiring hypothesis, which postulates that the formation of axons and dendrites effectively minimizes resource allocation while maintaining maximal information storage.

Sensory Processing

Early models on sensory processing understood within a theoretical framework are credited to Horace Barlow. Somewhat similar to the minimal wiring hypothesis described in the preceding section, Barlow understood the processing of the early sensory systems to be a form of efficient coding, where the neurons encoded information which minimized the number of spikes. Experimental and computational works have since supported this hypothesis in one form or another. For the example of visual processing, efficient coding is manifested in the forms of efficient spatial coding, color coding, temporal/motion coding, stereo coding, and combinations of them.

Further along the visual pathway, even the efficiently coded visual information is too much for the capacity of the information bottleneck, the visual attentional bottleneck. A subsequent theory has been developed on exogenous attentional selection of visual input information for further processing guided by a bottom-up saliency map in the primary visual cortex.

Current research in sensory processing is divided among a biophysical modelling of different subsystems and a more theoretical modelling of perception. Current models of perception have suggested that the brain performs some form of Bayesian inference and integration of different sensory information in generating our perception of the physical world.

Motor Control

Many models of the way the brain controls movement have been developed. This includes models of processing in the brain such as the cerebellum's role for error correction, skill learning in motor cortex and the basal ganglia, or the control of the vestibulo ocular reflex. This also includes many normative models, such as those of the Bayesian or optimal control flavor which are built on the idea that the brain efficiently solves its problems.

Memory and Synaptic Plasticity

Earlier models of memory are primarily based on the postulates of Hebbian learning. Biologically relevant models such as Hopfield net have been developed to address the properties of associative (also known as "content-addressable") style of memory that occur in biological systems. These attempts are primarily focusing on the formation of medium- and long-term memory, localizing in the hippocampus. Models of working memory, relying on theories of network oscillations and persistent activity, have been built to capture some features of the prefrontal cortex in context-related memory. Additional models look at the close relationship between the basal ganglia and the prefrontal cortex and how that contributes to working memory.

One of the major problems in neurophysiological memory is how it is maintained and changed through multiple time scales. Unstable synapses are easy to train but also prone to stochastic disruption. Stable synapses forget less easily, but they are also harder to consolidate. One recent computational hypothesis involves cascades of plasticity that allow synapses to function at multiple time scales. Stereochemically detailed models of the acetylcholine receptor-based synapse with the Monte Carlo method, working at the time scale of microseconds, have been built. It is likely that computational tools will contribute greatly to our understanding of how synapses function and change in relation to external stimulus in the coming decades.

Behaviors of Networks

Biological neurons are connected to each other in a complex, recurrent fashion. These connections are, unlike most artificial neural networks, sparse and usually specific. It is not known how information is transmitted through such sparsely connected networks, although specific areas of the brain, such as the Visual cortex, are understood in some detail. It is also unknown what the computational functions of these specific connectivity patterns are, if any.

The interactions of neurons in a small network can be often reduced to simple models such as the Ising model. The statistical mechanics of such simple systems are well-characterized theoretically. There has been some recent evidence that suggests that dynamics of arbitrary neuronal networks can be reduced to pairwise interactions. It is not known, however, whether such descriptive dynamics impart any important computational function. With the emergence of two-photon microscopy and calcium imaging, we now have powerful experimental methods with which to test the new theories regarding neuronal networks.

In some cases the complex interactions between inhibitory and excitatory neurons can be simplified using mean field theory, which gives rise to the population model of neural networks. While many neurotheorists prefer such models with reduced complexity, others argue that uncovering structural-functional relations depends on including as much neuronal and network structure as possible. Models of this type are typically built in large simulation platforms like GENESIS or NEURON. There have been some attempts to provide unified methods that bridge and integrate these levels of complexity.

Visual Attention, Identi ication and Categorization

Visual attention can be described as a set of mechanisms that limit some processing to a subset of incoming stimuli. Attentional mechanisms shape what we see and what we can act upon. They allow for concurrent selection of some (preferably, relevant) information and inhibition of other information. In order to have a more concrete specification of the mechanism underlying visual attention and the binding of features, a number of computational models have been proposed aiming to explain psychophysical findings. In general, all models postulate the existence of a saliency or priority map for registering the potentially interesting areas of the retinal input, and a gating mechanism for reducing the amount of incoming visual information, so that the limited computational resources of the brain can handle it. An example theory that is being extensively tested behaviorally and physiologically is the hypothesis of a bottom-up saliency map in the primary visual cortex. Computational neuroscience provides a mathematical framework for studying the mechanisms involved in brain function and allows complete simulation and prediction of neuropsychological syndromes.

Cognition, Discrimination and Learning

Computational modeling of higher cognitive functions has only recently begun. Experimental data comes primarily from single-unit recording in primates. The frontal lobe and parietal lobe function as integrators of information from multiple sensory modalities. There are some tentative ideas regarding how simple mutually inhibitory functional circuits in these areas may carry out biologically relevant computation.

The brain seems to be able to discriminate and adapt particularly well in certain contexts. For instance, human beings seem to have an enormous capacity for memorizing and recognizing faces. One of the key goals of computational neuroscience is to dissect how biological systems carry out these complex computations efficiently and potentially replicate these processes in building intelligent machines.

The brain's large-scale organizational principles are illuminated by many fields, including biology, psychology, and clinical practice. Integrative neuroscience attempts to consolidate these observations through unified descriptive models and databases of behavioral measures and recordings. These are the bases for some quantitative modeling of large-scale brain activity.

The Computational Representational Understanding of Mind (CRUM) is another attempt at modeling human cognition through simulated processes like acquired rule-based systems in decision making and the manipulation of visual representations in decision making.

Consciousness

One of the ultimate goals of psychology/neuroscience is to be able to explain the everyday experience of conscious life. Francis Crick, Giulio Tononi and Christof Koch made some attempts to formulate consistent frameworks for future work in neural correlates of consciousness (NCC), though much of the work in this field remains speculative. Specifically, Crick cautioned the field of neuroscience to not approach topics that are traditionally left to philosophy and religion.

Computational Clinical Neuroscience

Computational clinical neuroscience is a field that brings together experts in neuroscience, neurology, psychiatry, decision sciences and computational modeling to quantitatively define and investigate problems in neurological and psychiatric diseases, and to train scientists and clinicians that wish to apply these models to diagnosis and treatment.

Brain Computer Interface

Brain Computer Interface (BCI) technology is a powerful communication tool between users and systems. It does not require any external devices or muscle intervention to issue commands and complete the interaction. The research community has initially developed BCIs with biomedical applications in mind, leading to the generation of assistive devices. Brain computer interface (BCI) systems build a communication bridge between human brain and the external world eliminating the need for typical information delivery methods. They manage the sending of messages from human brains and decoding their silent thoughts. Thus they can help handicapped people to tell and write down their opinions and ideas via variety of methods such as in spelling applications, semantic categorization, or silent speech communication.

Machine Learning and Brain Computer Interfaces

Machine learning methods for Brain Computer Interface pursues architecture of two functions:

- Training function.
- Prediction function.

In supervised learning, model M is selected that can learn on the basis of either parametric or non-parametric tests and can map on a set of input and output data. In unsupervised learning, machine learns the specific structure in the input space on the basis of given learning example. Figure describes the architecture of machine learning methods for BCI.

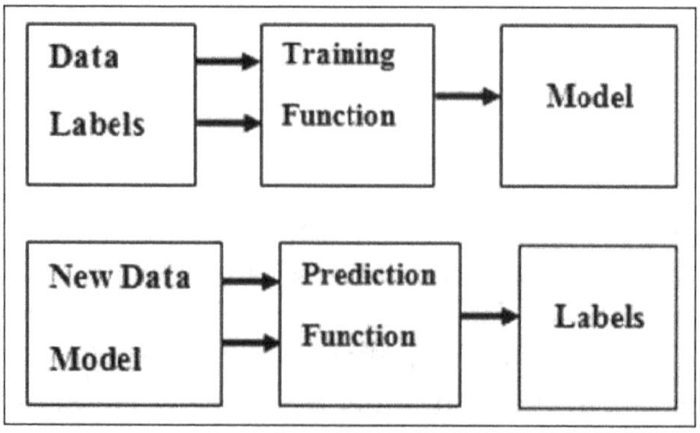

Architecture of Machine Learning Methods for BCI.

Background and Scope of Machine Learning Methods for BCI zProbabilistic methods are developed in machine learning which discover structures and patterns available in data. These methods then are applied to scientific real-time problems. In Neuroinformaitcs or in Computational Neuroscience, the study of how the information is processed by brain and how it is analyzed and interpreted are carried out. Neuroscientific experiments are conducted for that. The first thing to be done is to do the classification. In classification, based on external observations a rule is found. There are just two different classes available in the simplest case. Then the function is estimated from the given function class F, The test examples are understood to be generated from the similar probability distribution p(x, y). By minimizing the expected error (error is nothing but the risk), best function f can be obtained. When the underlying probability distribution is unknown, the risk cannot be minimized directly. Following is the formula of approximation of minimum risk.

To give condition is possible on the learning machine that can make sure the empirical risk will congregate to the expected risk asymptotically. The large deviation is possible for small sample size and over fitting may take place. Figure shows the illustration of over fitting.

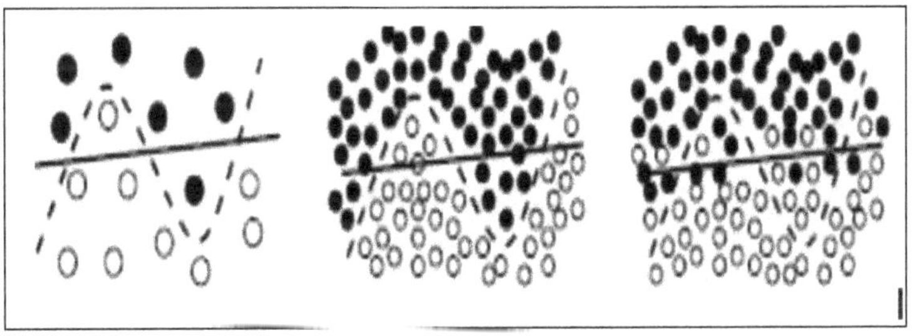

Illustration of over Fitting.

Linear Classifiers and Methods

Linear classifiers and methods are commonly used in Brain Computer Interfaces. Linear classification makes use of simple model. But the underlying probability distribution is not assumed properly and if outliers are present results obtained can go wrong and this happens in the analysis of Brain Computer Interfacing data. In Brain Computer Interfacing data k (k=1.......K) is measured first. Samples are Xk, where x is represents each sample point. To obtain linear hyper plane classifier, normal vector W is to be estimated and a threshold b is calculated by some optimization technique. Following is the equation of obtaining hyper plain.

A linear classifier is defined by a hyper plane's normal vector w and an offset b, i.e. the decision boundary is {x |wTx+b = o} (thick line). Each of the two half spaces defined by this hyper plane corresponds to one class, i.e. f (x) =sign (wTx+b). The margin of a linear classifier is the minimal distance of any training point to the hyper plane. In this case it is the distance between the dotted lines and the thick linc. There are various linear classification techniques like optimal linear classification.

Nonlinear Classifiers, Ahead of the Scope of Linear Classifiers

In particular, kernel related learning follows nonlinear classification. However, overall classification is nonlinear by nature in input space; feature space and input space are nonlinearly related. By using nonlinear mapping, algorithms in feature spaces use following plan.

In this equation, data x1, x2,.....,xn ∊ Rn. Rn is mapped with feature space F. So now the same algorithm is considered for F at the place of Rn. In this classification, first regression on F is calculated. Input data is mapped with the hypotheses space or with RBF bumps in radial basis networks or in neural networks or in boosting algorithm. Nevertheless learning in F can be simple if linear classifiers (a low complexity) are used. In figure somewhat complicated nonlinear decision surface (two dimensional classifications) is used to separate the classes in a feature space. For the separation hyper plane is needed. In this instance two things are controlled:

- Learning machine's algorithmic complexity by using feature space.
- Statistical complexity by using simple linear hyper plane figure represents two dimensional classifications.

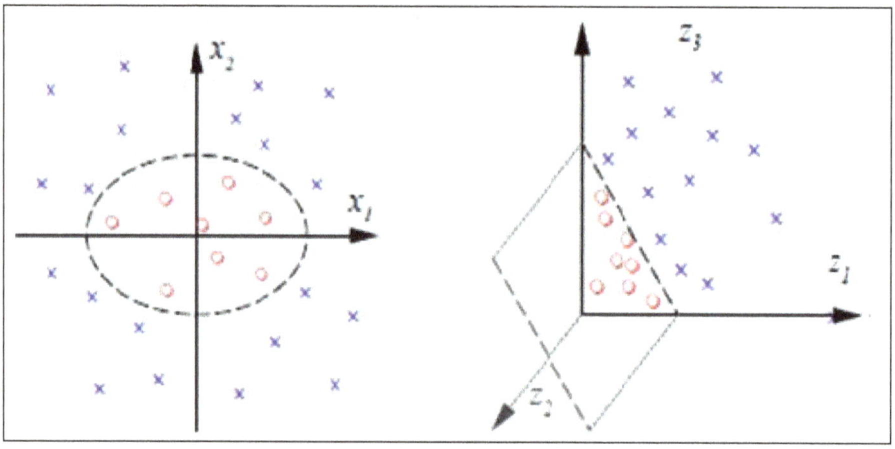

Two Dimensional Classification.

Computational Evolutionary Biology

Computational evolutionary biology (or computational evolution) is the study of evolutionary biology using computers. This makes it a fuzzy sub-discipline of computational biology, overlapping with bioinformatics and computational genomics. It is not evolutionary computation, evolutionary programming or evolutionary algorithms, all of which apply evolutionary principles to optimization problems outside biology.

The methods involved in computational evolutionary biology are diverse. They include:

Bioinformatics algorithms for clustering, tree-building, sequence alignment machine learning etc. that mine and analyze large-scale data. Uses include:

- Building phylogenetic trees to discover evolutionary relationships between species.
- Deciphering the evolutionary history of specific genes and traits.
- Studying the rate of evolution and selection pressures on particular genes.
- Address fundamental evolutionary questions from an empirical perspective.

Numerical simulations such as individual-based models, agent-based models, and spatially explicit models. These are used to:

- Explore evolutionary dynamics of species and their interactions.
- Predict evolutionary events such as speciation and extinction.
- Support mathematical models of complex phenomena, such as individual development and cooperation in groups.
- Link the predictions of idealized mathematical models with behavior of real systems, such as the evolution of bacterial drug resistance and long-term behaviour of epidemics.

Artificial life explores the foundations of biology as a computational phenomenon by creating algorithms that evolve freely within an environment specified by the programmer. Examples include Tierra and Avida, cellular automata and evolving neural nets.

References

- Computational-neuroscience, glossary: emotiv.com, Retrieved, 02, August 2020
- Machine-Learning-and-Computational-Neuroscience-A-Widespread-Move-towards-Brain-Computer-Interface, publication-325027971: researchgate.net, Retrieved, 12, March 2020
- What-is-computational-evolutionary-biology: quora.com, Retrieved, 06, May 2020
- Computational-Neuropsychiatry-of-Working-Memory-Disorders-in-Schizophrenia-The-Network-Connectivity-in-Prefrontal-Cortex-Data-and-Models, publication-238315124: researchgate.net, Retrieved 10, January 2020
- Computational-pharmacology-service, Computational-pharmacology-is-a-rapidly-data-from-many-diverse-sources: creative-biolabs.com, Retrieved 05, June 2020

3

Systems Biology: Modeling and Clustering

Systems biology is an approach in biomedical research which uses analysis and modeling of complex biological systems. It employs a holistic approach and focuses on the smaller pieces. This chapter has been written carefully to provide an easy understanding of systems biology as well as its scope and impacts.

Systems biology is concerned with the study of biological functions and mechanisms, underpinning inter- and intra-cellular dynamical networks, by means of signal- and system oriented approaches. Systems Biology's a well-established area of human activity that can be traced back at least to the writings of Aristotle around 350 B.C. The term "system" signifies here an entity that represents wholeness, which can be superficially divided (fractioned) into components (parts) but whose pivotal properties cannot be fully explained solely from the knowledge of the parts. The paradigms of life sciences were changing with time but the core fundamental concepts of systems biology are apparently resistant to these changes and remain pretty much the same today as they were in time of Aristotle.

Human Systems Conceptualized to Work as Machines

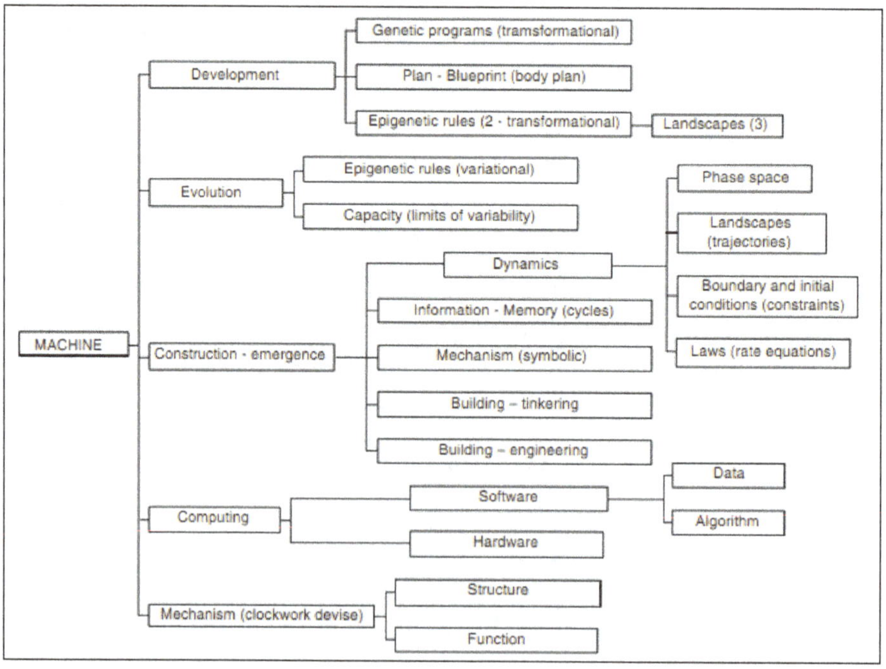

Computational and Mathematical Approaches and Tools

Our increased awareness that biological process is regulated by intracellular networks there has been recent focus on methods to target the networks with combinations of drugs or genetic perturbations rather than single points in the network. Targeted perturbations using small molecule inhibitors have very useful in revealing the complexity of circuits in cancer cells by uncovering regulatory feedback loops which can serve to bypass the perturbations. Perturbations of genomic expression levels using siRNA, shRNA, or miRNA show that the signaling networks are highly dynamic and capable of rewiring circuits within the cancer cells that then establish new context specific interactions that allows cells including tumor cells to persist in the presence of the perturbation. These adaptive responses likely limit the efficacy of many targeted therapeutics in use or in clinical trials. The emerging challenge that limits the significance of these efforts to functionally characterize cancer cells is how to integrate the large amount of divergent data sources into executable, predictive, system's level models that can be used to positively impact cancer treatment by determining fragility points in the networks that can be targeted for personalized cancer therapy.

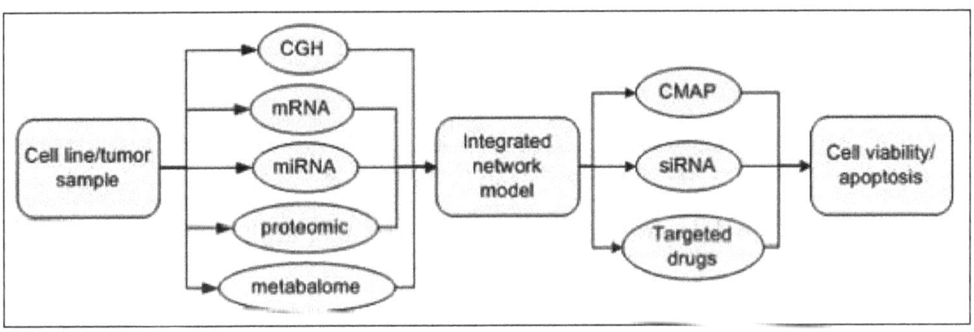

Scope of Systems Biology

System Biology essentially involves:

- Methods and technologies for monitoring cell-wide gene expression.
- Mathematical and computational analysis of gene expression data.
- Methods and technologies for monitoring cell-wide protein expression.
- Mathematical and computational analysis of protein expression data.
- Methods and technologies for identification of protein-protein interaction.
- Mathematical and computational analysis of protein-protein interaction data.
- Methods and technologies for identification of DNA-protein interaction.
- Mathematical and computational analysis of DNA-protein interaction data.
- Genetic networks.
- Mathematical methods for the identification of genctic regulatory networks.

- Modeling and Simulation of gene expression networks.
- Translation networks.
- Modeling and Simulation of protein expression networks.
- Methods and technologies for monitoring metabolic reaction networks.
- Mathematical and computational analysis of metabolic reaction networks.

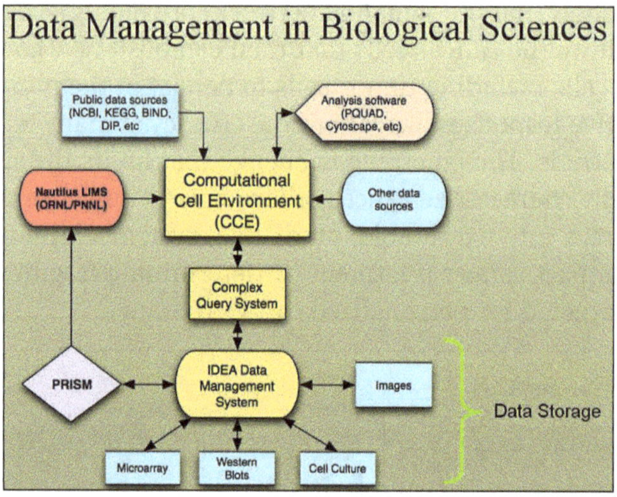

Data Management in Biological Science.

Tools and methods for data collection, analysis, storage, and sharing are being integrated.

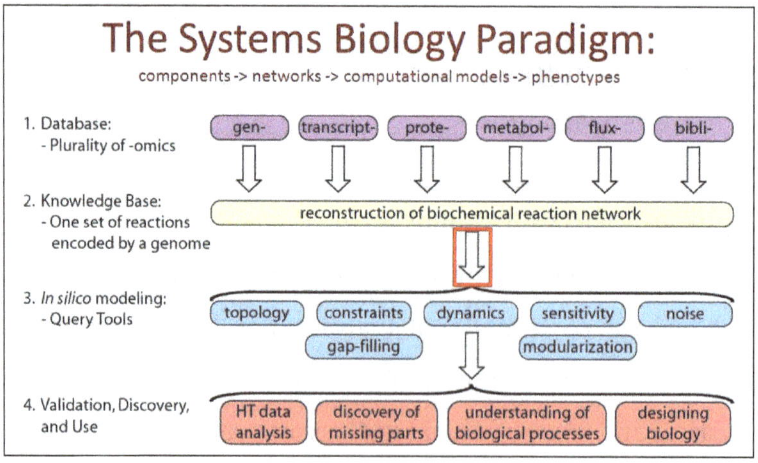

Cancer Systems Biology

Cancer is now appreciated as not only a highly heterogeneous pathology with respect to cell type and tissue origin but also as a disease involving dysregulation of multiple pathways governing fundamental cell processes such as death, proliferation, differentiation and migration. Thus, the activities of molecular networks that execute metabolic or cytoskeletal processes, or regulate these by signal transduction, are altered in a complex manner by diverse genetic mutations in concert with the environmental context. A major challenge therefore is how to develop actionable understanding of this multivariate dysregulation, with respect both to how it arises from diverse genetic mutations and to how it may be ameliorated by prospective treatments. While high-throughput experimental platform technologies ranging from genomic sequencing to transcriptomic, proteomic and metabolomic profiling are now commonly used for molecular level characterization of tumor cells and surrounding tissues, the resulting data sets defy straightforward intuitive interpretation with respect to potential therapeutic targets or the effects of perturbation. In this review article, we will discuss how significant advances can be obtained by applying computational modeling approaches to elucidate the pathways most critically involved in tumor formation and progression, impact of particular mutations on pathway operation, consequences of altered cell behavior in tissue environments and effects of molecular therapeutic.

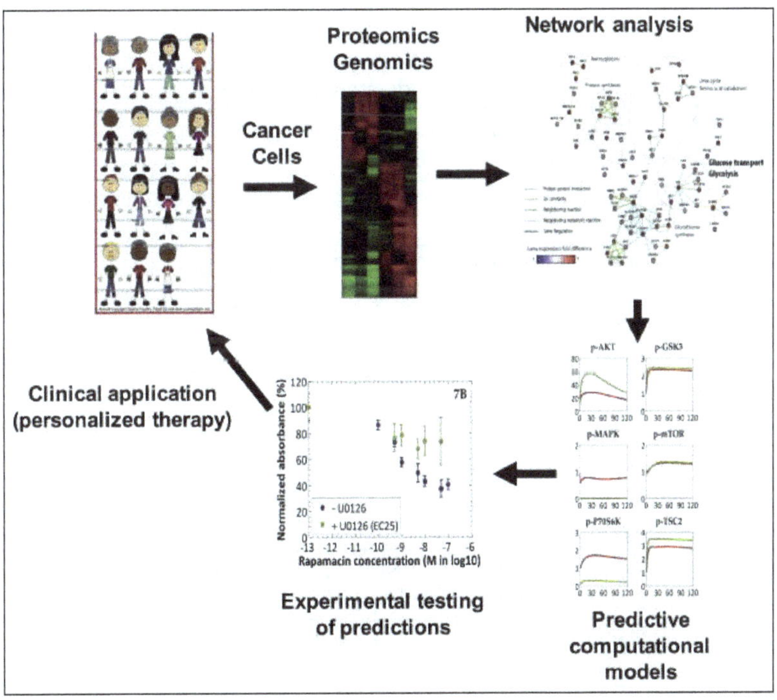

Cancer Systems Biology Resources

High-throughput technologies have been developed to enable comprehensive genomic analyses of mutations, rearrangements, copy number variations, and methylation at the cellular and tissue levels, as well as robust analysis of RNA and microRNA expression data, protein levels and metabolite levels. In parallel, predictive data-driven mathematical and computational methods have been developed to analyse the large amounts of data arising from the new analysis platforms. Unfortunately, none of the platforms alone is sufficient to capture the complexity of changes in the genome of a cancer cell or to discriminate optimally between cancer driver and passenger mutations. In addition, all methods generate noise and have their own intrinsic weaknesses, thereby hampering interpretation.

So far, the ability to integrate and interpret data across platforms has substantially lagged behind the ability to generate the data; however, integrative genomic studies can provide a new paradigm for the discovery of cancer aberrations and interactions, novel treatments, and resistance mechanisms. Large-scale projects have generated terabytes of 'omics' data that have been curated and made available to the international scientific community. To integrate the data and present them in a way that biological translations can be generated, large programs have developed novel computational tools and databases of interactions that are accessible to the larger community. Furthermore, independent of these large-scale projects, many online resources contain annotated high-throughput data, computing tools and visualization, databases of interaction and quantitative biological information.

Project
TCGA (The Cancer Genome Project): TCGA performs comprehensive genomic characterisation and analysis of multiple cancer types to drive the understanding of the molecular basis of cancer.
ICGC (International Cancer Genome Consortium): ICGC is focused on a comprehensive understanding of the genomic abnormalities in cancer.
SU2C (Stand Up to Cancer): SU2C aims to drive innovative cancer research through interdisciplinary science.
LINCS (Library of Integrated Network-based Cellular Signatures): LINCS aims to enhance a network-based understanding of biology, cataloging changes in gene expression and other cellular processes in response to perturbations.
Sanger Institute databases: The institute has developed a suite of databases to manage and interpret large-scale data.
cBioPortal for Cancer Genomics: CBIO is a platform to visualise, analyse and download large scale genomic data sets.
The Human Protein Atlas: The atlas aims to explore the human proteome systematically through antibody-based proteomics.
The Cancer proteome atlas: The atlas aims to provide a comprehensive resource for accessing, visualising, and analysing cancer functional proteomics.
TCGA data portal: A platform for researchers to search, download, and analyse data sets generated by TCGA.
Cytoscape: Cytoscope is an open source platform for visualising complex networks and integrating the networks with other data types.
Vcell: The Virtual cell is a computational platform to model and simulate cell biology.
COPASI: COPASI is a platform to simulate and analyse networks and their dynamics.
Netwalker: Netwalker is a platform to assist in functional analyses of large-scale genomics datasets focused on molecular networks.
SageBio: SageBio creates platforms to enable collaboration on data and data sharing. They also run challenges on complex biomedical problems.
STRING: String is a database of known and predicted physical and functional protein interactions.

GeneCards: GeneCards provides a database of human genes with comprehensive information on all known and predicted human genes.
Pathway Commons: Pathway Commons is a portal to access biological pathway information collected from public pathway databases.
DREAM (Dialogue for Reverse Engineering Assessments and Methods): DREAM aims to be a catalyser for the interaction between experiment and theory focused on cellular network inference and quantitative model building.
ICBP (Integrated Cancer Biology Program): ICBP develops and implements computational models of cancer processes.

The existing systems biology resources contain different types of data that are synergistic and complementary and form the spokes of the cancer systems biology wheel. On one side, TCGA, ICGC, and Stand Up to Cancer projects have focused on the collection of tumour samples and on the characterization of their molecular profiles together with data on patient outcomes. On the other side, CCLE, LINCS, and ICBP projects have focused on obtaining a detailed molecular characterization of the responses of several cancer cell lines to multiple perturbations. Online databases and computing resources contain detailed information on genes, proteins, and molecular interactions as well as online computing software and tools to visualize and analyze large datasets.

The data collection and computing efforts through many programmes have provided a rich resource for the implementation of cancer systems biology and have supported the progression of the field. Moreover, adapting new efforts, such as 'crowd sourcing' to large data analysis, have already provided an exciting and novel way to analyse large system-level biological data. Projects such as the Dialogue for Reverse Engineering Assessments and Methods (DREAM challenge), have made 'big data', collected through efforts such as the ICBP, available to thc international community of computational biologists and mathematicians.

Cancer Systems Biology Approaches

In spite of extensive research investigating and uncovering many important aspects of cancer biology, we are still far from an integrated understanding of how the genomic and epigenetic abnormalities that occur in cancer cells mediate their functional consequences. Many groups have assessed large clinical cohorts by measuring one molecular data type (such as gene expression, mutation, single-nucleotide polymorphism) in relation to a measurable outcome to derive correlations between a particular molecular event and disease subtypes. These observational studies are important, but they are insufficient to demonstrate causality. Furthermore, the interactions between genes, proteins, and metabolites are coordinated in intracellular and intercellular networks to enable effective interaction with the microenvironment and with all organ systems in our body, necessitating both the acquisition and integration of data across multiple levels. The cellular networks that are perturbed in cancer are, in general, the same cellular pathways used by normal cells to perccive and respond to the environment, although not always with the same interactions or spatio-temporal dynamics, owing to the rewiring of the networks as a result of molecular aberrations. Furthermore, in some cases, gene mutations or fusions can result in new functions that are not mediated by the parental molecule. These neomorphic functions engendered by genomic aberrations in cancer cells are particularly difficult to deconvolute and integrate into cellular pathways and networks. Because aberrations can occur at various levels and lead to changes in multiple parts of a signalling network, integrating multiple data types from the same tumour becomes necessary in order to derive a more global or 'systemic' understanding of the molecular drivers of the cancer.

The TCGA Research Network has developed perhaps the most comprehensive data set encompassing exon and whole-genome sequencing data combined with analysis of DNA copy number, mRNA, microRNA, promoter methylation and protein expression. In each tumour lineage examined, new observations have been reported. For example, the genomic analysis of high-grade serous ovarian cancer in the North American and European populations has revealed the existence of TP53 mutations across virtually all tumours and the dominant effect of DNA copy number aberrations, Furthermore, this analysis identified specific microRNA, mRNA, or methylation subtypes and the activation of the NOTCH and FOXM1 pathways as key features of ovarian carcinoma. Similar analyses in breast cancer increased our understanding of previously identified subsets, highlighted the recurrence of mutations in TP53, PIK3CA, and GATA3 genes as well as specific mutations within subtypes, such as PIK3CA in luminal tumours, and led to the identification of new subtypes of breast cancer that were not obvious from previous analyses. Interestingly, integrated analyses identified dominant signalling pathways driven by HER2 or EGFR activity. based on the downstream phosphorylation of the EGFR HER2 signaling network stressing the differential treatment of patients within this subgroup moving forward. In endometrial, colon, and rectal cancer, hypermutated tumours seemed to be a result of microsatellite instability, with a new type of instability driven by pole gene mutations resulting in ultramutated tumors. Integrated analyses also revealed MYC-directed activation in aggressive colorectal carcinoma. The use of this type of analysis in clear-cell renal cell carcinoma identified cellular oxygen sensing, chromatin remodeling/histone methylation, and metabolic shifts in the tricarboxylic acid (TCA) cycle as key processes in this pathology. Of note, integrated analyses across different cancer types highlighted some striking similarities in terms of molecular characteristics for basal-like breast cancer, high-grade serous ovarian cancer, and serous endometrial cancer. Systems approaches integrating data across multiple diseases in a 'pan cancer' effort aimed at improving our knowledge of the molecular pathogenesis of cancer whereby data is analyzed across tumor types to identify molecular characteristics common to a range of cancer types versus disease sub-types. The greatest opportunities associated with such analysis are only beginning to emerge. The multiplatform data sets on highly characterized patient samples provide the information necessary for the development of robust and predictive models. However, the importance of an iterative approach wherein hypotheses and predictions arising from the modelling are refined and constrained by experimental evaluation must be emphasized. In addition to such large consortial efforts, several groups have employed systems approaches using novel computational methods to integrate different data types to elicit novel principles. Carro et al., using gene expression and network approaches, identified key transcriptional factors, such as CEBP and Stat3, as master regulators of the mesenchymal transformation in glioblastoma. Importantly, they demonstrated that the expression of these transcription factors could predict poor clinical outcome. In another approach the integration of protein and phosphoproteomic interactions with transcription factor-DNA binding data led to the discovery of novel features of cellular responses to stress and growth factors. Similarly, systems-level analyses of proteomic and metabolomic data revealed novel regulatory functions between these two important cellular networks that would not have been achieved otherwise. Furthermore, the integration, through network analysis, of mRNA expression data with proteomic, phosphoproteomic and clinical outcomes data, allowed the identification of key regulators of EGFR signalling and correlated these events with patient survival. Whole-genome sequencing and proteomics were also used to identify new driver genes in endometrial cancer and to identify a novel functional association between ARID1A and PI3K pathway activity.

Systems Approaches for Tissue Complexity

In any tissue in the body, there is extensive interaction between cells within the organ and the microenvironment, giving rise to an additional layer of complexity whereby multiple networks must work together cohesively. These same interactions occur between the tumour and its microenvironment, regulating the process of tumorigenesis. Many of these interactions are contextual, with markedly different effects at different stages of tumour development. For example, it is clear that TGFβ in the tumour microenvironment can mediate both tumour growth and inhibition. However, an understanding of the mechanisms underlying these disparate processes has remained elusive. This complex functionality may in fact only be solved via a systems biology approach. The molecular aberrations in cancer can significantly alter the normal dynamics of (neo) vascularisation and hypoxia response.. Furthermore, the role of the microenvironment in tumour dormancy and evading therapeutic interventions that allows oestrogen receptor-positive breast cancer to recur years later remains elusive.

Owing to the complexity of the tumour microenvironment interactions, many other aspects of cancer biology have not been explained by traditional scientific approaches and can benefit by the integrated analysis supported by system biology. Imaging of cells and their interactions with the extracellular environment can provide the high-quality data needed for the development of predictive models. To distinguish a stromal signature from a tumour signature and to understand the influence of different (metastatic) microenvironments, or how stromal expression changes can influence tumour progression, several strategies have been used. An extensive analysis of gene-expression patterns in primary and metastatic tumours, applying novel statistical analytical methods, led to the discovery of genes that were reprogrammed in brain metastases, but not in other metastatic sites. This reprogramming event was a function of the local microenvironment, likely regulated by promoter methylation. Moreover, Finak et al. purified tumour stromal cells by laser capture microdissection and compared gene-expression data of the tumour stroma to healthy stroma, and developed a stromal signature with clear prognostic impact independent of tumour subtypes. Thus, using systems-based methods, several groups have shown that the stroma has an important interplay with the tumour and has clear effects on disease progression and possibly also on effective treatment options. Developing computational models to fit the complex dynamics of the cellular microenvironment can provide potential hypotheses to be tested experimentally and aid in the interpretation of the results. This iterative approach provides the basis for a truly integrative modelling and experimental platform. Several experimental approaches are well adapted to feed data for this high-complexity integrative platform.

Three-dimensional culture systems provide a potential approach to more realistically mimic the in vivo situation. Moreover, microfluidic systems, where tissues and even multiple different cell lineages can be studied in relatively high-throughput (micro) models, provide another experimental approach to support the modelling of the interactions between the tumour and microenvironment. These approaches have demonstrated critical environmental and spatial aspects of cellular responses within the tumour microenviroment. These in vitro approaches, therefore, have the potential to provide tractable experimental systems to develop hypotheses to test in vivo and also to experimentally test mathematical models developed from in vivo studies. Integrating in vitro and in vivo modeling approaches can play a major role in understanding how spatial orientation and interactions with the microenvironment can affect diverse tumour behaviour.

Response to (targeted) treatment and tumor progression are intrinsically linked to immunological pathways. At a single platform level, the complex interplay from the multitude of players cannot be completely captured. Using a pathway recognition algorithm and multilevel data from TCGA, Kristensen et al. showed that tumors can be stratified by their immune signatures. This stratification is shown to be relevant for patient survival and targeted treatment response. Further underlining the importance of the relationship between the tumor and the microenvironment, multiscale analysis showed the tumor can recreate its immune environment throughout cancer progression.

Several investigators have developed modelling approaches to study the relationship between intracellular pathways and the behaviour of cells at the tissue level. Recently, some researchers modelled myc and p53 pathways during the proliferation of lymphoma cells in the context of surrounding tissue, growth factors, and angiogenesis. In addition, models considering the effect of cell division rates as well as interactions between cancer cells and the microenvironment predicted that the order in which targeted and chemotherapy approaches are delivered, is important to optimizing benefit whereby the proliferation rates and migration capabilities of different cell types could alter the tissue level response to therapy. Critically, these models will help to predict how effects at small intervals within the cell relate to long-term effects observed at the tissue or whole body level. Mathematical models of the spatio-temporal heterogeneity of the extracellular matrix have provided unexpected insights into how, for example, the urokinase plasminogen activation system impacts on cell invasion. The aforementioned studies primarily used 'omics' data derived from analyses of tumour lysates without the consideration of spatial orientation, one of the key aspects of the tumour microenvironment is spatial constraints. To incorporate spatial information, imaging of tumour cells and their environment is of paramount importance, particularly in the context of angiogenesis. The differential imaging of blood vessels and cell types using fluorescent probes as well as intravital imaging of tumours in animal models and cell cultures are techniques that provide the data necessary to develop 'tissue-level' computational models that can incorporate all these facets. For example, two predictive models for angiogenesis and cell-cell interactions have been developed, by incorporating imaging data, in order to study in vivo aspects of the tumour biology and drug response, respectively. Furthermore, Haeno et al. used a dissemination dynamics model (including well-annotated patient data) to predict the type and sequential order of treatments most likely to benefit patients with pancreatic cancer. This model can have important clinical implications for this disease that has a notoriously poor prognosis. Importantly, Kim and colleagues highlighted the applicability and individual benefits of different types of modelling in combination with in vitro and patient data to study spatio temporal drug distribution within the tumor to optimize drug selection and delivery.

Systems Approaches for Cell Heterogeneity

Cancer stem-like cells, also known as tumour-initiating cells, have been intensely studied in recent years. These studies have been driven in part by the discovery that the presence of cancer stem-like cells likely contributes to drug resistance and poor outcome. Interestingly, characteristics previously attributed to epithelial-mesenchymal transition (EMT), including EMT markers and mesenchymal morphology, can contribute to or be associated with a cancer stem cell-like state and drug resistance. The ability to transition in and out of an EMT-like state is critical for a cell to initiate and complete the metastatic process. Systems-based analysis of prostate cancer identified a molecular signature associated with patient outcome. Moreover, the results from this study highlighted an extensive plasticity of basal cells, supporting a model in which cells of origin can generate distinct molecular subtypes of prostate cancer.

Different models of the spatiotemporal dynamics of stem cells in both healthy cells and tumour cells have been published to evaluate normal olfactory epithelium, bowel crypts, and solid tumours in an attempt to understand the regulatory strategies of cell renewal. These models give rise to the hypothesis that stem cells are not cell types as such, but rather a behavioural state imposed on the cell by feedback mechanisms.

Such mechanisms are still in place in tumours, although they are susceptible to perturbations, including those induced by treatment. Changes in tumour size can alter cell proliferation and death by disturbing the feedback mechanisms and shift the balance towards cell renewal. These models could help in predicting which approaches might successfully shrink the tumour and benefit patients.

Intratumoural and intertumoural heterogeneity contribute to patient prognosis and predicts response to therapy. Intratumoural heterogeneity may represent the greatest challenge to deliver effective personalized cancer therapy. Tyson et al. developed novel automated tools to capture and quantitate multiplexed imaging features to develop predictive models of the heterogeneity of response of individual cells to perturbations and were able to show heterogeneous cell fates upon exposure to anti-proliferative drugs. In another study, Yankeelov et al., used quantitative tumour imaging methods and developed a model of interactions between cancer cells, stroma and immune cells, vascularisation and the extracellular matrix, that was able to predict treatment response and tumour progression. These modeling approaches have been very useful in enabling us to understand cellular heterogeneity and response to perturbations and lay the foundation of how we can use such information to develop novel methods to treat tumors which are inherently heterogeneous.

Systems Approaches for Targeted Therapy

One of the continual challenges in the use of targeted therapy is the low response rates observed in the clinical setting. In many cancer systems biology efforts, novel computational and mathematical methods are used to integrate and analyse patients' molecular data to identify optimal markers and targets. Recent efforts focusing on colorectal and liver cancers have integrated whole-genome expression data with patient's outcomes to classify patients and their therapy responses according to gene-expression patterns. Using unbiased analysis of these datasets, the investigators identified specific expression patterns and pathways that were associated to specific mutations in the tumour, and which altered drug response in the cancer cells. Furthermore, extensive computational analysis of a large metabolic network revealed key regulatory nodes, such as oxidative phosphorylation, glycolysis and citric acid cycle that could serve as optimal targets for therapy. Several groups have also integrated phosphoproteomic time-course data of cell line responses to growth factors and perturbations in epithelial tumours, developing predictive data-driven models to identify rational targets and combinations of drugs for effective therapy. In diseases with limited therapeutic options (such as pancreatic cancer) an extensive systems biology approach integrating gene-expression data from primary pancreatic ductal adenocarcinomas and from human or murine pancreatic adenocarcinoma cell lines, revealed novel classifiers, predicted potential subtype-specific therapies, and identified non-responders.

Beyond pancreatic cancer, an area of intense research in terms of system biology has been breast cancer. Much work has in fact focused on molecular phenotypes and drug responses in breast cancer. These efforts used several systems-level approaches, including large 'omics' data acquisition, analysis of patients and responses to clinically approved drugs as well as large panels of cell line-based assays. Computational modeling and analysis of perturbation screens using non-coding RNA have also proven useful in defining network topologies and predicting combinations of targets based on mechanistic functions of subnetworks.

Approaches for Drug Resistance

Drug resistance is one of the major challenges to effective cancer-treatments. Integration of thousands of 'omics' datapoints acquired from sensitive and resistant tumours into a systems-based therapeutic strategy has been shown to be a powerful tool for tackling resistance. For example., models of dynamic changes in signalling networks of receptor tyrosine kinases (RTK) families have facilitated the classification of RTKs and their network activation and led to the identification of points of intervention to delay or overcome drug resistance. Furthermore, Komurov et al., used a novel network-based analysis of gene expression and proteomics coupled to ErbB2-positive patient survival data, and showed that in breast tumours with acquired resistance to lapatinib, the drug was still able to block EGFR/ErbB2 signalling, but that upregulation of glucose metabolism, unfolded protein response, and endoplasmic reticulum (ER) stress pathways mitigated the ability of lapatinib to induce cell death, suggesting that coordinated targeting of metabolic networks and signalling networks has the potential to improve patient outcomes. Beckman et al. illustrated that the potential effects on drug response and resistance acquisition induced by single cell heterogeneity and cellular dynamics can be mathematically modeled and, therefore, should be included in personalized treatment strategies. They demonstrated that a strategy targeting all cancer subpopulations including the precursors of resistant clones, rather than treatment targeting the clone predominantly present which therefore initially leads to the largest reduction in tumour size, may be more effective. As such counterintuitive strategies may be of advantage for patient outcome. Lastly, integrating patient data with murine models has proven a powerful systems approach to identify mechanisms to overcome resistance to androgen therapy in prostate cancers. All these different studies highlight how using systems biology approaches can improve targeted therapy and help to overcome drug resistance.

With the aim of improving effective combinatorial drug prediction in cancer, mathematical modelling combined with high-density time-dependent measurements showed that modulation of oncogenic pathways through sequential application of drug combinations is possible and can render cancer cells much more susceptible to the drugs, thus significantly increasing the efficacy of treatment. Context-specific metabolic modelling algorithms demonstrated feasibility in predicting drug targets and phenotypic tumour response, including off-target effects, making this an effective systems approach for the development of new (combinatorial) drugs therapy. For example, STK38

(Serine-threonine kinase 38,) was identified as an upstream regulator of MYC activity through a model of network dynamics(modulator inference by network dynamics (MINDy) Algorithm) aimed at predicting transcription factor modulators, and experimental manipulation of STK38 demonstrated inhibition of tumour growth in MYC-dependent tumors in vivo; and thus a potential novel means to target these tumours.

Cancer systems biology has made an impact on drug discovery and development from the identification of targets up to the clinical trial stage. Computational analysis of the ErbB family network has revealed ErbB3 as a key node of ligand-induced ErbB2 activation. These results facilitated the engineering of a bispecific ErbB3 antibody, MM-111, that inhibits ligand-induced receptor activation in an ErbB2-overexpressed environment and as such bypasses the tumour ErbB3 escape route that is often activated in drug resistance. Computational modelling together with pharmacodynamic and tumour growth algorithms aimed at identifying optimal drug combinations (with MM-111) to prevent drug resistance. Of note, the MM-111 antibody, a drug that would not likely have been developed without a systems biology approach, is currently in clinical development.

Impact of Cancer Systems Biology

There has been an explosion in the quantity of available experimental data from highthroughput technologies, such as genomic sequencing, transcriptomics, metabolomics, and proteomics. Additionally, targeted experiments on a smaller scale have provided important information about complex interactions within and between cells. Systems analyses and predictive modeling are necessary to integrate across these datasets that span different length and time scales to convert them into knowledge. Systems Biology brought together cancer biologists and oncologists with computer scientists, mathematicians, and engineers to tackle important cancer questions including the identification of tumor master transcriptional regulators, the role of microRNAs in cancer progression and metastasis, the development of drug resistance in heterogeneous tumor contexts, and optimization of combination therapies for cancer treatment. It is envisioned that the successes of cancer systems biology will encourage investigators to bring these tools to bear on emerging and difficult cancer questions that will require a systems approach to fully comprehend.

Complex phenotypes emerge from the interactions of thousands of bio-molecules that are organized into complexes and interacting with one another as one functional module which in turn, emerges as functional networks in health and disease. Omics approaches collect data on changes for all genes or other macromolecules in the cells and statistical analysis attempts to uncover the functional modules that perform the functions that characterize higher levels of biological organization. Systems biology attempts to analyze the study of individual genes/proteins and to integrate them into one functional unit. Cancer cells exhibit an altered /defective genetic and epigenetic landscape formed by altered complexes and network modules arising in different parts of tumor tissues that sustain autonomous cell behavior ultimately promoting tumor growth. Understanding of tumor biology must address not only molecular but also, and more importantly, tumor cell heterogeneity, by considering cancer tissue genetic and epigenetic networks.

Brain Cancer

The most common types of brain cancers are astrocytomas, which are malignant tumors that arise from a class of brain cells called astrocytes. The tumors can be classified by their level of malignancy on a grade scale from I – IV, with Grade I representing benign, non-cancerous tumors, and Grade IV (high-grade astrocytomas) representing highly invasive, very aggressive malignant tumors. These Grade IV astrocytomas are also referred to as glioblastoma multiforme, or GBM for short. As the disease progresses, the overall survival rate decreases dramatically from 8-10 years for patients diagnosed with Grade II tumors to 9-12 months for those diagnosed with GBM. Unfortunately, the majority (~70%) of low- and medium-grade astrocytomas go undetected only to emerge, without clinical warning, as GBM. In addition to the obvious tragedy this represents for patients and their family members, the dismally low detection rate of lower grade astrocytomas also means that researchers seeking to understand the progression of this disease are severely handicapped, lacking the crucial tissue samples from the stages that lie between health and advanced disease.

To address this challenge and gain insight into the mechanisms underlying the GBM development, researchers from across the country, including Burak Kutlu, Rhishikesh Bargaje and Lee Hood at ISB, used a group of mice that had been genetically modified to begin developing astrocytomas upon exposure to a specific chemical. Because the researchers would be able to control the disease initiating event, they could monitor both the transition from health to disease and the progression of the tumor from low-grade to high. This proved to be a powerful model for studying GBM development. Focusing in on three major networks that had previously been shown to be affected in GBM patients, the researchers were able to determine the exact order, from tumor initiation through progression to full-blown GBM, in which the networks malfunctioned. More specifically, they now know which of the three networks to examine for identifying biomarkers of early disease.

Lung Cancer

For a patient, determining whether a growth is benign can be as important, and as challenging, as detecting cancer. Often, diagnosis of a growth requires an invasive and costly surgical biopsy. Lung nodules are growths that, if malignant, may lead to lung cancer. A whopping 80 percent of patients who are diagnosed with having lung nodules endure invasive biopsies that turn out to be unnecessary. Researchers at ISB and Integrated Diagnostics (Indi is an ISB spinout) designed a blood test consisting of a panel of 13 biomarkers that will assist in identifying benign lung nodules. It has a 90% negative predictive value (NPV), which means that if the test calls the nodule benign (i.e. the result is negative), 90% of the time the test is correct. Indi currently is in the process of commercializing the diagnostics test, which ultimately will save patients from unnecessary — and expensive — invasive procedures.

Modeling in Systems Biology

Describe the Molecules and Interactions in a Mathematical Model

To be able to do a quantitative model of a biochemical network, a system has to be defined where molecular concentrations are the variables and their interactions are described explicitly by mathematical functions. These functions depend on the type of interactions to be familiar with common 'translations' for reactions, transcription, etc. into equations, with the ultimate goal of a complete quantitative model for the network.

Estimate Parameter Values for the Model

The mathematical description includes a number of parameters defining for example reaction rates. Different values of these parameters can result in different behaviours of the model. Hence it is crucial to estimate the parameter values that are relevant for a specific biological network. One possibility is to experimentally measure the parameter for a specific reaction, which will result in an optimal single estimated value for the parameter.

A potential drawback is that it is hard to do such measurements within a biological organism, and if it is measured elsewhere that specific condition might lead to a different value compared to within the organism. Another approach is reverse engineering, where model parameters are estimated by fitting model output to available experimental data.

Analyse the Dynamical Behaviour

A final step in a modelling approach is to analyse the behaviour of the defined model. Many molecular networks and modules show very high robustness. This should then also be accounted for in the model and can be tested by a sensitivity analysis, where the change of behaviour is tested when parameters are perturbed. Also, the model can be tested for perturbations where molecules or interactions are removed from the system, which then can be compared with knockout experiments, or provide biological predictions from the model. The analysis can provide feedback into the previous three steps improving the description and knowledge of the biological system.

Mathematical Modeling in Systems Biology

Biological systems are inherently complex, and many of their emergent properties result from the interplay of numerous molecular components. Moreover, biochemical reactions often obey nonlinear reaction kinetics—that is, an increase in the amount of the starting material of the reaction does not necessarily lead to a proportional increase in the amount of the reaction product. Finally, other complexities, such as cell structure and compartmentalization or random (i.e., stochastic) effects, also often result in unexpected behavior of the entire system. Mathematical models that take these factors into consideration allow researchers to capture the features of complex biological systems and to understand how biological systems respond to external or internal signals and perturbations, such as different growth or development conditions or stress triggered by agents such as alcohol. Mathematical models have the big advantage of being amenable to computer simulations. Models describing biological systems generally are too complex to be solved analytically ("manually") and therefore typically are solved numerically— that is, using computers to solve the mathematical equations that help predict the response of a biological system. With the availability of computer-based techniques for solving mathematical equations, the response of a biological system to different conditions can be relatively easily simulated in silico once a mathematical model is available. These computer simulations (socalled "dry experiments") in many cases require much lower investment and much less time compared with the typically more time-consuming and expensive biological experiments (sometimes referred to as "wet experiments").

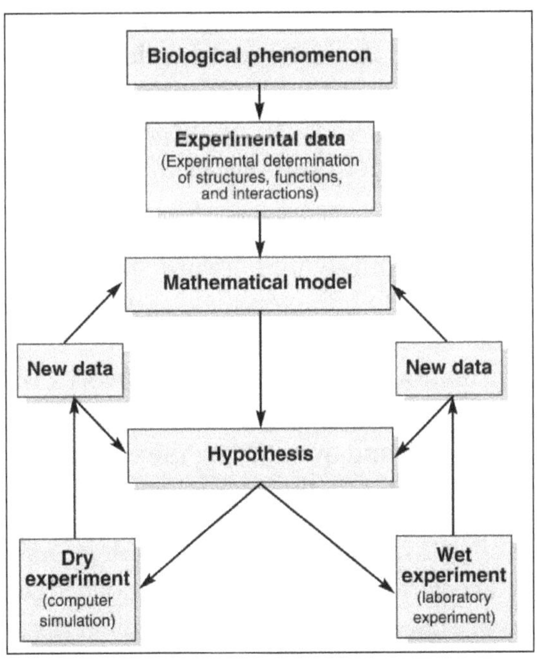

Network in System Biology

The development in molecular biology tools and technique of high-throughput techniques such as microarray technology availability of completely sequenced genomes and the and two-hybrid systems have allowed genome-wide studies of biological processes that arise from complex interaction between genetic entities, such as proteins, RNAs and DNA . These cellular entities interact with one another according to their functional roles. For instance, protein-protein interactions, related to cellular communication by signal transduction, activate or repress the transcription of genes, changing the molecular composition of the cell. These separate modules are 'building blocks' of the biological systems, working together to shape the phenotypic pattern of the cells and organisms. Mapping out all these interconnected modules helps to understand and model their topological and dynamical properties.

The most critical networks for controlling cellular systems are those related to transcription, protein-protein interaction and metabolism. Despite the great diversity of networks in cell biology, they share several global properties. For instance, most networks within the cell are characterized by (i) a scale-free degree distribution (also called power-law), (ii) a small average shortest path length between any two nodes (small-world), (iii) a disassortative nature, (iv) a modular organization and (v) a structural and dynamical robustness.

In the era of post-genomic biology, scientists have proposed the new and innovative idea that cells can be understood in terms of their network structure. To define the network structure of a cell, network biology researchers use data generated by experimental methods, including high-through put proteomic, genomic, and metabolomic data, as well as computational capabilities to identify and map cell signaling networks. It has been shown that by identifying the most highly connected proteins, the overall network topology can be inferred. These interaction networks can be defined in terms of the protein complexes within cells and their relationships to each other.

Viewing cells in terms of their underlying network structure is a very powerful concept. All networks share common characteristics, and mathematical treatments have been developed to understand their structure and how they can be regulated. Thus, organizing biological information in the context of networks is fundamental to applying systems-level approaches to understanding biological function. By defining protein complexes as dynamic structures and quantifying the extent and quality of interactions of the partners, information relating to their function can be inferred.

Revolutions in biotechnology and information technology have produced enormous amounts of data and are accelerating the process of knowledge discovery of biological systems. These advances are changing the way biomedical research, development and applications are conducted. Clinical data complements biological data, enabling detailed descriptions of both healthy and diseased states, as well as disease progression and response to therapies. The availability of data representing various biological states, processes and their time dependencies enables the study of biological systems at various levels of organization, from molecules to organism and even up to the population level. Multiple sources of data support a rapidly growing body of biomedical knowledge;

however, our ability to analyze and interpret this data lags far behind data generation and storage capacity. Mathematical and computational models are increasingly used to help interpret biomedical data produced by high-throughput genomics and proteomics projects. The application of advanced computer models enabling the simulation of complex biological processes generates hypotheses and suggests experiments. Computational models are set to exploit the wealth of data stored on biomedical databases through text mining and knowledge discovery approaches.

Modeling is the human activity consisting of representing, manipulating and communicating real-world daily life objects. As one can easily realize, there are many ways to observe an object or, equivalently, there are many different observers for the same object. Any observer has 'different views' of the same object, i.e. 'there is no omniscient observer with special access to the truth'. Each different observer collects data and generates hypothesis that are consistent with the data. This logical process is called 'abduction'. Abduction is not infallible, though; with respect to a scientific unknown, we are all blind.

A system is a collection of interrelated objects. For example, a biological system could be a collection of different cellular compartments (e.g. cell types) specialized for a specific biological function (e.g. white and red blood cells have very different commitments). An object is some elemental unit upon which observation can be made but whose internal structure is either unknown or does not exist. The choice of the elemental unit defines the representation scale of the system. A model is a description of a system in terms of constitutive objects and the relationships among them, where the description itself is, in general, decodable or interpretable by humans.

Generally speaking, a system is an unknown 'black box' (S) which, under a specif ic external stimulus (input E) produces a response (output R). Using this general definition, one can identify three primary scientific uses of models: (i) synthesis or knowledge discovery; to use the knowledge of inputs E and outputs R to infer system characteristics; (ii) analysis and prediction; to use the knowledge of the parts and their stimuli (i.e. the inputs E) to account for the observed response (i.e. the output R) and eventually, to predict response to different stimuli. (iii) Instrumentation or device; to design an 'alternative system' (i.e. hardware or software), able to reproduce the input–output relationship with the best possible adherence to the studied system.

Secondary uses of models account for conceptual frameworks to design new experiments, methods to summarize or synthesize large quantities of data, tools to discover relationships among objects.

The mathematical methods used in modeling biological systems vary according to different steps of the process. We focus on the mathematical representation of the system. However, other important steps in the modeling processes are parameters fitting and model selection. We will not analyze the mathematical methods in those two important aspects as these would require separate review papers. Methods for parameters fitting refer to wide area of mathematical optimization, whereas methods for model selection mainly use statistical techniques. On top of these, sensitivity analysis and phase– space analysis of the models may be required.

Models of Systems

Not all scientific models are expressed in a precise, numerical and quantitative way. Actually, one can identify four different types of models: verbal models, conceptual or diagrammatic models, physical models and formal models.

Verbal Models

In verbal models the system is described in words. These models, based on observations, usually evidence in a simple way the objects and relations among the objects in the system. A verbal model is a rough and sometime ambiguous qualitative representation of the knowledge of the system. These kinds of models are used in the first approach to the analysis of biological system.

Conceptual or Diagrammatic Models

In conceptual or diagrammatic models the system is described by a graphical representation of the objects and the relationships describing the underlying dynamical processes. To develop these kind of models the understanding of the available data needs to be sufficient to have a detailed (even if not exhaustive) idea of the objects (or entities) and relations. A conceptual model (CM) represents 'concepts' (objects or entities) and relationships between them. In computer science, CM are also referred to as domain models. A CM is expressly independent from the design and free from implementation concerns. The aim of a CM is to convey the meaning of terms and concepts used by 'domain experts' to rationalize the problem and to find relationships among the different concepts. The CM aims to clarify the meaning of the usually ambiguous terms to minimize as much as possible problems arising from different interpretations of terms and concepts. If a 'domain ontology' is available, then the meaning of the variety of terms used should be linked to it. Once the domain concepts have been modeled, the model becomes a stable basis for subsequent development of applications in that specific domain. Furthermore, the concepts behind the conceptual model can be mapped to physical design or implementation constructs using either manual or automated code generation approaches.

A conceptual model can be described using various notations, such as Unified Modeling Language (UML), Object Modeling Technique (OMT) for object modeling, or Information Engineering or Integration Definition for Information Modeling (IDEF1X) for Entity Relationship Modeling. In UML notation, the conceptual model is often described with a class diagram in which classes represent concepts, associations represent relationships between concepts and role-type of an association represents role types taken by instances of the modeled concepts in various situations.

Physical Models

In physical models the representation is done using a mock-up of a real system or object (like a scale model of an aircraft or of a ship). These types of models are mostly of interest for engineers. They are widely used when the properties of the system are almost 'scale-invariant', i.e. independent from the size of the physical model built to represent the real system to produce smaller-scale prototypes.

Formal Models

Formal models represent the knowledge of the system using mathematical structures. The mathematical representation of the model depends on the knowledge of the system, on some modeling choices (for instance, the spatial scale of representation) and the aim of the modeling process. There are a large variety of mathematical/computational methods that can be used and the selection of the proper one follows rules that are often matters of experience. At a first glance, there are few questions one may ask to address the choice of the proper mathematical/computational method. Those questions are mainly related to the description of the system with respect to its different parts or components, the physical variables space and time, the type of relations between objects and the object representation perse. In systems biology, a system is viewed as an assembly of different parts or compartments (i.e. organs) with different functions. In this case, 'Compartment models' are widely used and each compartment may pick a different mathematical representation. Models can also represent physical variables in different ways. Besides, the model may or not consider the evolution of the system with respect to time (dynamic versus static models). Time can be treated as continuous or discrete variable (time-continuous versus time-discrete models). Likewise, spatial distribution of objects in each compartment may or not be relevant (spatially-heterogeneous versus homogeneous models). Finally, similar objects may be treated as individuals or taken in bulk (particle models versus population models). In the first case, individual objects are identified by a unique state or by a large, but finite number of states (one-state particle versus finite-state automata). Lastly, the relations between objects can be described as deterministic or stochastic rules (deterministic versus stochastic models).

According to the different modeling choices, one can get single versus multi-compartments models, including transport, evolutionary differential equations versus algebraic equations or spatial partial differential equation, differential equations versus difference equations, ordinary differential equations (ODE) versus partial differential equations (PDE), kinetic methods, agent-based methods (ABM) or cellular automata (CA) versus ODE or PDE; deterministic methods (ODE or PDE, etc.) versus stochastic methods (stochastic ODE and PDE).

Statistical and Artificial Intelligence-based Models

A statistical model is a formalization of the relationships between variables (i.e. object's measurable characteristics) in the form of mathematical equations, the only difference with the mathematical models described above is that in statistics, all variables and/or parameters of the model include a level of uncertainty. When the relationship between two objects is too complex to be easily guessed, one can resort to probabilistic measures and statistical or artificial intelligence methods to reproduce the response relationship. In these kinds of models, the detailed analysis of the system components is usually ignored because; the objective of the model is limited to reproduce the system stimulus/response relation. Examples of this approach are the lumped models using equivalent circuits, neural network, etc.

The Modeling Process

The modeling process consists of the following steps: (i) model implementation consisting in describing by a formal language the objects/relationships identified in the system under study using a mathematical structure and/or a computer code; (ii) use the model to forecast the system behavior and (iii) evaluate the model adherence to reality by matching predictions with available data.

To find a good model is an issue. Modeling is a hard problem in itself and failure is not a rare event. The modeling procedure is a process in itself that follows a semi-formal set of rules. The methodology lean on four macro steps: (i) understand the problem, i.e. clearly define the questions one asks to the model; (ii) devise a plan for solving the problem, i.e. define a series of steps to be put in practice to find an adequate model of the system under investigations. This step includes knowledge and data acquisition from field experts and literature, model structure, model hypothesis, conceptual model, choosing the appropriate mathematical formalism, solving the formal model, get the results, check model results matching to available data and so on; (iii) execute the plan, i.e. perform the steps in (ii) and (iv) check the correctness of the answer and eventually refine the model. This last point is a major test to evaluate the hypothesis formulated when setting the model.

For these models, the classical description of modeling process is shown in figure. It is worth to mention that the schema illustrated in figure does not have the pretense to be the most general one: it is a general approach that can be used in the analysis and predictions models.

Model Objectives

A proper definition of the model objectives is a fundamental step as it implies a certain level of comprehension of the problem. The reason for building a model should by clear and a proper clarification of the objective must answer to major questions: (i) what is the system to be modeled; (ii) what are the major questions to be addressed by the model; (iii) How good must the model be and to what it will be compared with? (iv) How we will analyze and use the model output? All these questions need to be clarified before we proceed in searching the current knowledge on the system we wish to model, as some informations may be more relevant than the others.

Current Knowledge

A second crucial step in the modeling process is to collect the knowledge on the system under investigation. This is conducted by consulting the scientific literature, including experimental reports and/or discussing with field-experts. In the biomedical field, data sets of literature record (e.g. Pubmed) can highly facilitate the task of browsing the vast amount of information available nowadays. In this respect, methods of data mining and data extraction may be very useful.

Model Structure

A model is a representation of a real system and has its own structure. For the sake of manageability the model structure should include, of all the acquired knowledge, only those considered to be relevant for the purpose of the study ('Realism'); the level of details of the model results should also be determined a priori ('Precision'); finally, a model can be general, that is applicable to other similar systems or specific to the system of interest ('Generality'). Realism, Precision and Generality are competing properties. Each of these properties trades-off against the other two. Deciding a model structure is to find a proper balance between those competing properties, which satisfy the model objectives. Decision on the model structure is crucial for defining the model hypothesis, diagrammatic model construction and mathematical formulation.

Hypotheses

The next step in the modeling process is to translate objectives and current knowledge we wish to include in the model in a list of specific working hypothesis. These are usually verbal statements, but could also be quantitative relationships. Working hypotheses are the basis of the model we are going to build and model results will depend on them. In doing the cycle refinement of model, the starting hypothesis should be critically, repeatedly analyzed.

Conceptual Model

The conceptual model is a graphical representation of the relevant system knowledge and model objectives that have been identified in the hypotheses. In the conceptual model compartments, objects and relations will be described in a diagram where the set of objects are fully clarified and relations bounded.

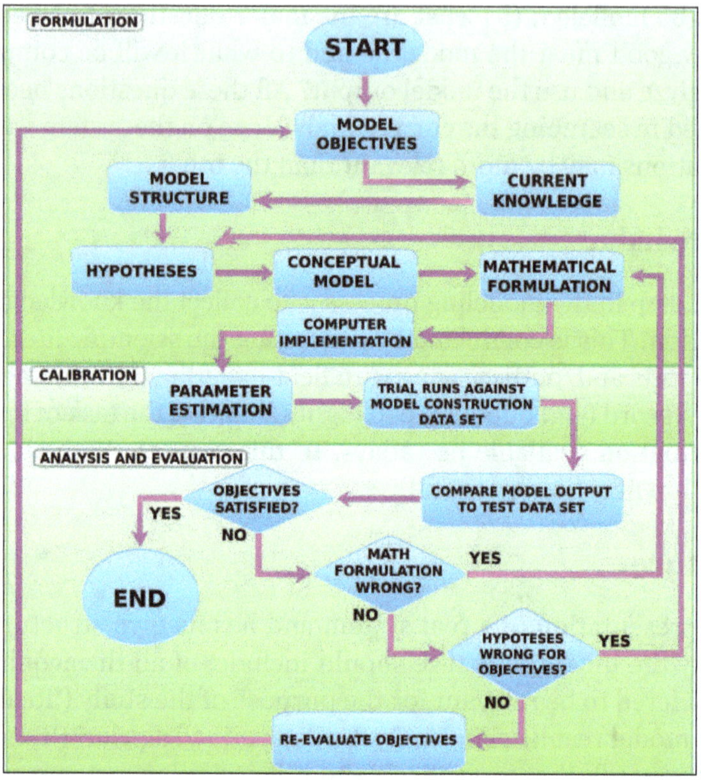

Figure shows the description of the modeling process. The top part of the figure refers to the formulation of the model, i.e. identification of the model objectives; collection of the current knowledge about the biological system under investigation; choosing the most appropriate model structure to satisfy the model objectives; translate objectives and knowledge into model hypotheses; draw a conceptual model; identify mathematical technique and develop the formal model. Central part of the figure refers to calibration i.e. estimating and fitting model parameters. Finally, the bottom of the figure refers to analysis and evaluation, i.e. comparison of the model results against experimental data sets and analysis of discrepancies.

Mathematical Formulation

This is usually the trickiest part of the modeling process, requiring the choice of a mathematical structure, which is appropriate for the model objectives and is able to describe in quantitative form the hypotheses. This step of the process requires a certain level of mathematical sophistication and, more importantly, it requires defining vague concepts and loose relations in strict mathematical terms. Noticeably, model objectives play an important role because; a detailed description of the biological system may turn out to be useless if not required by the model objectives.

Choosing mathematical formulations is a mapping of the model into the mathematical domain to obtain a formal model. A good formal model must be a compromise between the competing properties of any model (Realism, Precision and Generality) and should take into account some specificity of the mathematical domain. Accordingly, we can identify the major properties of formal models as follows: 'relevance', capturing the essential properties of the phenomenon; 'computability', transfer model hypothesis into a mathematical/computational infrastructure that can be solved to give the desired results with the required precision; 'understandability', offering a conceptual framework for thinking about the scientific domain; 'extendibility', allowing the inclusion of additional real-world objects in the same mathematical scheme. Taking into account the biological complexity, a very detailed model of a biological system may work out to be unsuitable to automatic resolution, i.e. not computable (for instance, too many equations to be solved or too many unknown parameters to be estimated); in contrast, a too simplistic model may not be able to account for the complexity of the biological system of interest.

A good formal model must be understandable by domain experts so that they could use it for their own quantitative reasoning. Finally, as biology is a fast growing science, extensibility is an important characteristic for biological models. When new objects and relations in the system are derived from laboratory experiments, it should be easy to extend the model with minor changes in the mathematical structure.

In most cases models are equipped with just one mathematical structure of those mentioned previously. However, there are also 'hybrid' models where different mathematical structures are used in combination but also models which add to a detailed description of the system of interest other types of models (e.g. BBM), which mimic the effects of other systems interlinked with the one under investigations.

From Mathematical Formulation to the Numerical Solution

Only very simple models can be analyzed analytically (i.e. by algebraic derivation of the system properties). In most cases the model is either directly implemented as a computer code (i.e. the algorithm-like in ABM) or equations must be solved numerically. Even if there are a variety of methods for solving equations, transferring equations into computer code is a possible source of error and appropriate method to avoid errors due to numerical instabilities need to be carefully chosen. Models of biological systems can involve scores of dynamic variables and thousands of parameters, especially when spatial processes are investigated. In this respect, checking the computer results to match available data is not a trivial exercise.

Parameter Estimation and Tuning

Once the model formulation has been translated on a computer, a further step is necessary before a simulation can be run. Complex models usually contain many parameters whose numerical value must be determined. In biology this may not be an easy task as numerical values of parameters, if directly measurable, are often known with a large uncertainty. Methods for parameter fitting are mainly based on optimization techniques that minimize the difference between real experimental data and model output. There are a variety of techniques ranging from stochastic methods and gradient descent methods. In some cases, specific experiments may be required for estimating parameters. However, a model can contain parameters that cannot be estimated a priori. In these cases one can have an a posteriori estimation. This means that the values of the parameters can be estimated by another model made ad hoc just to fit the data available.

Model Validation and Cyclic Refinement

Comparison of model results or simulations is the final part of the modeling process. The general goal of a model is to reproduce data from observation or from experiments (descriptive models) or to predict the result of new observations or experiments (predictive models). Obviously, results need to be validated according to model's objectives. In some cases, a qualitative agreement between model results and experimental data is adequate, in other cases, a quantitative agreement is necessary. It is a common practice for model validation to require the model results to be validated against independent data sets. Model results that fail to fit the experimental data set notwithstanding changes in the model-free parameters, suggest a further model refinement. In this respect, in the way back and forth between model refinement and data validation one can discover interesting properties of the system of interest. The process itself leads to discovering of new knowledge.

Models in Biology: Scales and Complexity

Any natural phenomena can be observed at different scales thus, in describing the phenomenon by conceptual and quantitative models one needs to choose to appropriate scale to describe the experimental data available.

However, in almost all complex natural phenomena there are aspects that cannot be even observed at just one scale of description (either temporal or spatial). To study these specific facets of reality, multiscale models that represent objects and relationships on different levels of abstraction are required.

Choosing a scale depends on which aspects of the phenomena, from 'micro' to 'macro', one is interested to analyze. In physics this is already a well-defined approach that originates from different research areas, and the distinction between different scales is based upon the characteristic lengths of objects and the characteristic time of the phenomena under investigation. For instance, microphysics refers to areas of physics that study phenomena, which take place on the microscopic scale (length scales <1 mm), such as: molecular physics, atomic physics, nuclear physics and particle physics.

In the life sciences the definition of a scale is a bit more ambiguous. A basic unit available for defining a scale is the 'cell' with no regard to its physical dimension. Starting from this, one can define different scales: the 'sub-cellular' or 'intra-cellular' scale, the 'cellular', mesoscopic or 'inter-cellular scale the 'macroscopic scale' and the 'populations scale'.

Models developed at sub-cellular scale deal with the evolution of the physical and biochemical state of a single cell. This scale involves genes, proteins and signals in cell nucleus and surface, which regulate the evolution of the cell and any signaling processing operations of the cells enabling cell crosstalk. Modeling the overall activity of a single cell is a very hard problem as many biological details of this activity are unknown.

Biologists and modelers have joined forces to develop and use mathematical and computer science techniques in modeling sub-cellular phenomena. Interested readers can find plenty of references in the scientific literature.

In the cellular scale, one is interested in describing the evolution of a system consisting of a large number of different interacting cells and molecules. Cell interactions are regulated by signals emitted and received by cells through complex recognition processes. Cellular scale is thus highly connected with the sub-cellular scale but, modeling at this scale, one may forget the details of single cell models and consider them as BBM. The areas of mathematical methods and tools involved in this description refer to statistical mechanics, cellular automata, lattice gas and other similar approaches.

The 'macroscopic scale' include tissues, organs (i.e. a collection of tissues joined in structural unit to serve a common function), systems (i.e. a group of organs working together to perform a certain task) and organism.

In this scale, one is interested in describing the dynamical behavior of observable quantities, in most cases, the concentrations of various entities (cells or molecules). Tissues are usually described using techniques originating from physical continuous systems, i.e. ordinary or partial differential equations or moments of kinetic equations. In describing organs, a model is required to describe both the main tissue and the sporadic tissues but also, and most importantly, the biological function. To model a system, one is required to consider a network of organs that perform a specific task. Depending on the modeling goal the model of a biological system can be arranged with different levels of details. Organs can be fully described in their components or simply as BBM performing a given task. Connections between organs (like lymphatic vessels) can be described physically (dynamical description of the fluid motion in the vessels) or simply considering the flux and the time required to move portions of fluid from different organs, i.e. though law of transport.

Finally, in the population scale, one is interested in describing the dynamics of the populations with respect to one or more characteristics. Models of epidemics or population controls are well known models at this scale. Population dynamics is extremely complex because the effects of all previously mentioned scales and the effects of the environment on a single organism can modify the overall dynamic of the populations. In this class of models, a single organism can or cannot be described in detail, according to the size of the populations one is required to describe. In both cases, changes of the major characteristics of a single organism must be taken into account. For example, to describe the response of a population to large-scale vaccinations (as required in influenza epidemics), one does not describe in detail any single organism but it may be required to consider the age structure of the population and the effects of environments. At variance for small size populations, like modeling the effect of a new vaccine for a small trial, a sufficiently detailed description of the organism and the effect of the vaccine on each organism should be required. A variety of different techniques are available for these classes of models. In the former case, one uses both, ordinary or stochastic differential equations to describe the populations dynamics or agent-based models (simple agents representing a single organism) to study the resulting complex phenomena. In the latter case, a more detailed description of the organism is required and the population dynamics may eventually be extracted from the dynamic response of each organism.

Complexity and Multiscale Models

Living organisms are complex systems. Using a 'classical' definition, a complex system is a system composed of different interconnected parts that, as a whole, exhibit one or more properties which do not obviously arise from the properties of the individual parts. System complexity may be either a 'disorganized complexity' or an 'organized complexity'. In the former case complexity arises from a very large number of parts, whereas in the latter case, complexity is intrinsic to the system, eventually with a limited number of parts, and its connections rules.

In living organisms both situations occur. A living organism is formed by a collection of different parts which are, each of them, organized complex systems. Cells, organs, systems of the human body are each of the complex systems.

As an example, the immune system is one of the very complex one where complexity arises both from a very large number of parts (organs), constituents (cells and molecules) and rules hierarchically connecting different scales of the parts.

Models including many scales of a phenomenon are now requested both for knowledge discovery and drug discovery. In life sciences not only an entire living organism, but also parts of the organism are too complex to be represented in a single, precise, multiscale model. The resulting model would certainly not be computable. Thus, one is forced to break the conceptual model in a set of models describing only part of the phenomenon (like a single organ, or a definite scale) and connect their outcomes. To link models at different scales is a not an easy task. Phenomena occurring at different scales have usually different characteristic time scales and models' output should be properly fitted.

Tools and Applications

Whether we investigate the growth and interactions of an entire population, the evolution of DNA sequences, the inheritance of traits, the spread of disease or the immune system response to a pathogen, biological systems are marked by change and adaptation. Even when they appear to be constant and stable, it is often the result of a balance of tendencies pushing the systems in different directions. The choice of the mathematical approach depends on the biological system one would like to model. In this section, we sketch several applications of mathematical techniques that have been effective in reproducing and in providing new insights of a particular biological problem.

Due to their incredible complexity, models that deal with an entire biological system are, to date, very few and actually incomplete. Instead there are several mathematical models that act toward single or group of components of a biological system.

Tools for Bioinformatics and Systems Biology

Signaling network has a key role in cellular physiology and therefore, it has been widely studied in several organisms. This is due mainly because all cells interact and respond to the environment in which they live. The bad news is that such networks are very complex due to their combinatorial explosion nature. For this reason, frameworks for mapping signal-transduction networks that avoids the combinatorial explosion in some way are particularly needed. A framework for mapping, visualization and automatic model creation of signal-transduction networks is presented, along with an example of its use to compile the, presently, most comprehensive map of the yeast MAP kinase network.

It is well known that the explosion of data originated from biology has made it increasingly important to provide metadata alongside the core data itself. The concept of metadata derives from cart catalogs and libraries by describing the contents and context of data files, the quality of the original data/files is greatly increased. This metadata may comprise domain-specific information as described by minimal information checklists meant to enable accurate data reuse or may be ontological in nature, specifying more precisely the kind of entities under consideration.

Next generation sequence analysis has become an important task both in laboratory and clinical settings. SeqAlto is a new algorithm for read alignment. It is about to 10 faster than existing algorithms, while retaining high accuracy and the ability to align reads with large (up to 50 bp) indels.

Proteins execute and coordinate cellular functions by interacting with other biomolecules. Among these interactions, protein–protein (including peptidemediated), protein–DNA and protein–RNA interactions cover a wide range of critical processes and cellular functions. Multi-VORFFIP is a tool to predict protein-, peptide-, DNA- and RNA-binding sites in proteins. One of its features is the web interface to facilitate the use of the method and analysis of predictions to non-expert end-users.

Applications

Immunology

The role of mathematical modeling in immunology, one of the most complex fields in biology, were recognized early, beginning from the 1960s and the 1970s. Since then, mathematical models have been used in various domains of immunology. One of the major issues in vaccine and other immunologic approaches' research is the testing of the relevant biological variables when each experiment lasts ≥ 1 year. One clear example is the scheduling of prolonged vaccinations. It is desirable to reduce as much as possible the number of vaccine administrations, e.g. to reduce the risk of side effects in humans. The authors describe the use of a mathematical model based on ABM that faithfully reproduce in silico the behavior of a cancer-preventive vaccination, suggesting a possible optimized vaccine schedule and highlighting certain critical issues. In particular, although vaccinations could be reduced in numbers without sacrificing efficacy, the intensity of early vaccinations was a key determinant of long-term tumor prevention needed for predictive utility in the model. Moreover, long-term studies confirmed predictions of in silico modeling in which an immune plateau phase, once reached, could be maintained with a reduced number of vaccinations; revealing that the accuracy of mathematical modeling of early immune responses is critical. This key example shows that an integrated in vivo-in silico approach could improve both mathematical and biological models of cancer immune-prevention. A model for keloid formation triggered by virus, their malignant effects and immune system competition have been described using a mathematical model developed by kinetic theory of active particles.

A mathematical model to analyze the co-stimulatory effects of anti-CD137 monoclonal antibody (mAb) for the melanoma treatment upon synergistic adoptive transfer of activated OT-1 T cells. The reported in vivo experiments show that a single administration of anti-CD137 mAb plus activated OT1 T cells is sufficient to completely reject the B16-OVA, whereas single components or not activated OT1 T cells have no success. The in silico experiments performed with the presented computational model show very good agreement with their in vivo counterpart. As many aspects of CD137 molecule biology are still not fully understood and the investigating of these aspects requires many difficult and expensive wet lab experiments, the model is a good candidate for becoming a predictive tool.

Circulatory System

Circulatory system represents a biological system made of organs that passes nutrients and other components to and from cells in the body to help fight diseases, stabilize body temperature and pH, and to maintain homeostasis. Diseases associated with this system, i.e. cardiovascular diseases, have a major impact in Western countries. Mathematical models and numerical simulations of cardiovascular system have been presented and have been demonstrated to provide help in understanding both their dynamics and possible interventions. In a previous study the authors provide a general overview of mathematical representation of vascular geometries extracted from medical images, the modeling blood rheology and the complex multilayer structure of the vascular tissue, and its possible pathologies and the mechanical and chemical interaction between blood and vascular walls.

Population Dynamics

With the integration of the difference equation model with an individual-based stochastic simulation model for malaria in humans, they compare the effects of insecticide-treated nets (ITNs) and indoor residual spraying (IRS) in reducing malaria transmission, prevalence and incidence. They conclude showing that ITNs are more effective than IRS in reducing transmission and prevalence, proving also that the combination of both interventions is more effective than either intervention alone.

Drug Efficacy

Molecular biology is the branch of biology that deals with the molecular basis of biological activity. This field share knowledge with other areas of biology and chemistry, i.e. genetics and biochemistry. Molecular biology provides the understanding of the interactions between the various systems of a cell, including the interactions between the different types of DNA, RNA and protein biosynthesis.

Clustering in Systems Biology

Clustering is the grouping of a particular set of objects based on their characteristics, aggregating them according to their similarities. Regarding to data mining, this methodology partitions the data implementing a specific join algorithm, most suitable for the desired information analysis.

This clustering analysis allows an object not to be part of a cluster, or strictly belong to it, calling this type of grouping hard partitioning. In the other hand, soft partitioning states that every object belongs to a cluster in a determined degree. More specific divisions can be possible to create like objects belonging to multiple clusters, to force an object to participate in only one cluster or even construct hierarchical trees on group relationships.

There are several different ways to implement this partitioning, based on distinct models. Distinct algorithms are applied to each model, differentiating its properties and results. These models are distinguished by their organization and type of relationship between them. The most important ones are:

- Centralized: Each cluster is represented by a single vector mean, and a object value is compared to these mean values.

- Distributed: The cluster is built using statistical distributions.
- Connectivity: The connectivity on these models is based on a distance function between elements.
- Group: Algorithms have only group information.
- Graph: Cluster organization and relationship between members is defined by a graph linked structure.
- Density: Members of the cluster are grouped by regions where observations are dense and similar.

Broadly speaking, clustering can be divided into two subgroups:

- Hard Clustering: In hard clustering, each data point either belongs to a cluster completely or not. For example, in the above example each customer is put into one group out of the 10 groups.
- Soft Clustering: In soft clustering, instead of putting each data point into a separate cluster, a probability or likelihood of that data point to be in those clusters is assigned. For example, from the above scenario each costumer is assigned a probability to be in either of 10 clusters of the retail store.

Clustering is the process of making a group of abstract objects into classes of similar objects:

- A cluster of data objects can be treated as one group.
- While doing cluster analysis, we first partition the set of data into groups based on data similarity and then assign the labels to the groups.
- The main advantage of clustering over classification is that, it is adaptable to changes and helps single out useful features that distinguish different groups.

Applications of Cluster Analysis

- Clustering analysis is broadly used in many applications such as market research, pattern recognition, data analysis, and image processing.
- Clustering can also help marketers discover distinct groups in their customer base. And they can characterize their customer groups based on the purchasing patterns.
- In the field of biology, it can be used to derive plant and animal taxonomies, categorize genes with similar functionalities and gain insight into structures inherent to populations.
- Clustering also helps in identification of areas of similar land use in an earth observation database. It also helps in the identification of groups of houses in a city according to house type, value, and geographic location.

- Clustering also helps in classifying documents on the web for information discovery.
- Clustering is also used in outlier detection applications such as detection of credit card fraud.
- As a data mining function, cluster analysis serves as a tool to gain insight into the distribution of data to observe characteristics of each cluster.

Requirements of Clustering in Data Mining

The following points throw light on why clustering is required in data mining:

- Scalability – We need highly scalable clustering algorithms to deal with large databases.
- Ability to deal with different kinds of attributes – Algorithms should be capable to be applied on any kind of data such as interval-based (numerical) data, categorical, and binary data.
- Discovery of clusters with attribute shape – The clustering algorithm should be capable of detecting clusters of arbitrary shape. They should not be bounded to only distance measures that tend to find spherical cluster of small sizes.
- High dimensionality – The clustering algorithm should not only be able to handle low-dimensional data but also the high dimensional space.
- Ability to deal with noisy data – Databases contain noisy, missing or erroneous data. Some algorithms are sensitive to such data and may lead to poor quality clusters.
- Interpretability – The clustering results should be interpretable, comprehensible, and usable.

Clustering Methods

Clustering methods can be classified into the following categories:

- Partitioning Method.
- Hierarchical Method.
- Density-based Method.
- Grid-Based Method.
- Model-Based Method.
- Constraint-based Method.

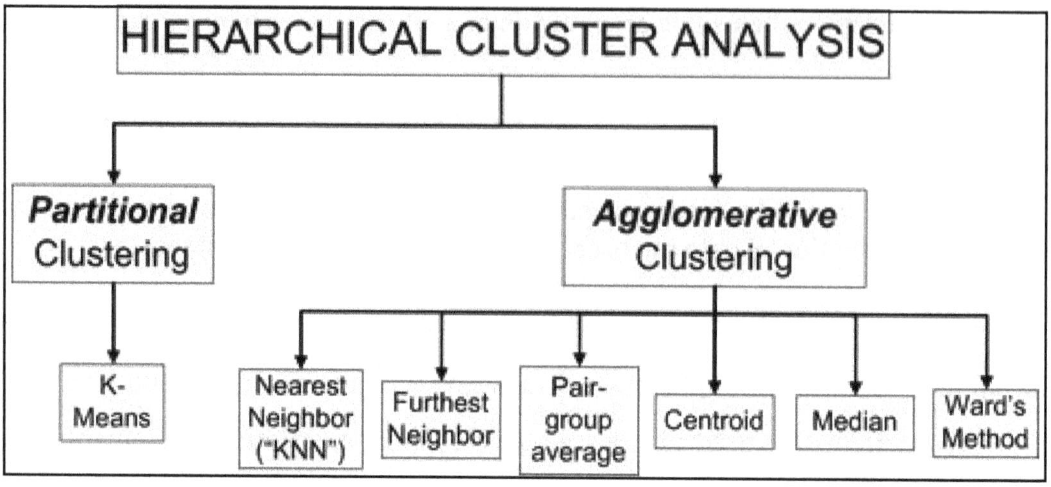

Partitioning Method

Suppose we are given a database of 'n' objects and the partitioning method constructs 'k' partition of data. Each partition will represent a cluster and k ≤ n. It means that it will classify the data into k groups, which satisfy the following requirements:

- Each group contains at least one object.
- Each object must belong to exactly one group.
- For a given number of partitions (say k), the partitioning method will create an initial partitioning.
- Then it uses the iterative relocation technique to improve the partitioning by moving objects from one group to other.

Hierarchical Methods

This method creates a hierarchical decomposition of the given set of data objects. We can classify hierarchical methods on the basis of how the hierarchical decomposition is formed. There are two approaches here:

- Agglomerative Approach.
- Divisive Approach.

Agglomerative Approach

This approach is also known as the bottom-up approach. In this, we start with each object forming a separate group. It keeps on merging the objects or groups that are close to one another. It keep on doing so until all of the groups are merged into one or until the termination condition holds.

Divisive Approach

This approach is also known as the top-down approach. In this, we start with all of the objects in the same cluster. In the continuous iteration, a cluster is split up into smaller clusters. It is down until each object in one cluster or the termination condition holds. This method is rigid, i.e., once a merging or splitting is done, it can never be undone.

Understanding Gene Expression

Gene expression is the process by which the genetic code of a gene is converted into a functional product, such as a protein for the production of cell structures. Gene regulatory network, biological network, transcription network, gene co-expression network, etc. are a few regulatory mechanisms through which gene expression is controlled. Gene expression and its various aspects are covered in detail in this chapter.

Gene Expression

Gene expression is the process by which information from a gene is used in the synthesis of a functional gene product. These products are often proteins, but in non-protein coding genes such as transfer RNA (tRNA) or small nuclear RNA (snRNA) genes, the product is a functional RNA. The process of gene expression is used by all known life—eukaryotes (including multicellular organisms), prokaryotes (bacteria and archaea), and utilized by viruses—to generate the macromolecular machinery for life.

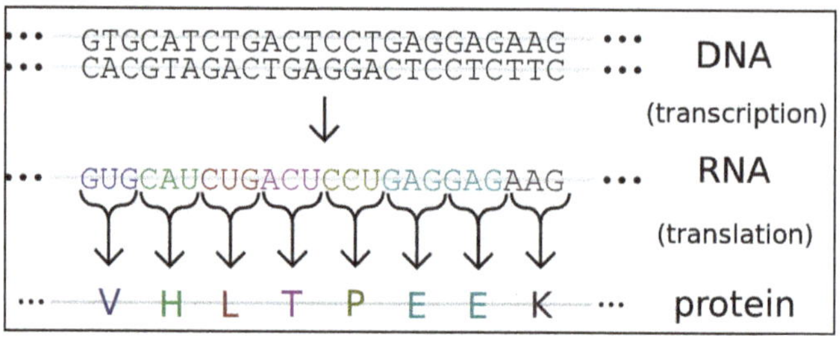

Genes are expressed by being transcribed into RNA, and this transcript may then be translated into protein.

Several steps in the gene expression process may be modulated, including the transcription, RNA splicing, translation, and post-translational modification of a protein. Gene regulation gives the cell control over structure and function, and is the basis for cellular differentiation, morphogenesis and the versatility and adaptability of any organism. Gene regulation may also serve as a substrate for evolutionary change, since control of the timing, location, and amount of gene expression can have a profound effect on the functions (actions) of the gene in a cell or in a multicellular organism.

In genetics, gene expression is the most fundamental level at which the genotype gives rise to the phenotype, i.e. observable trait. The genetic code stored in DNA is "interpreted" by gene expression, and the properties of the expression give rise to the organism's phenotype. Such phenotypes are often expressed by the synthesis of proteins that control the organism's shape, or that act as enzymes catalysing specific metabolic pathways characterising the organism. Regulation of gene expression is thus critical to an organism's development.

Mechanism

Transcription

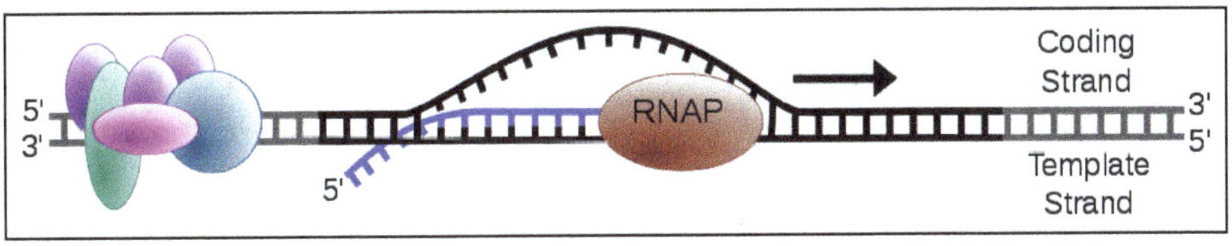

The process of transcription is carried out by RNA polymerase (RNAP), which uses DNA (black) as a template and produces RNA (blue).

A gene is a stretch of DNA that encodes information. Genomic DNA consists of two antiparallel and reverse complementary strands, each having 5' and 3' ends. With respect to a gene, the two strands may be labeled the "template strand," which serves as a blueprint for the production of an RNA transcript, and the "coding strand," which includes the DNA version of the transcript sequence. (Perhaps surprisingly, the "coding strand" is not physically involved in the coding process because it is the "template strand" that is read during transcription.)

The production of the RNA copy of the DNA is called transcription, and is performed in the nucleus by RNA polymerase, which adds one RNA nucleotide at a time to a growing RNA strand as per the complementarity law of the bases. This RNA is complementary to the template 3' → 5' DNA strand, which is itself complementary to the coding 5' → 3' DNA strand. Therefore, the resulting 5' → 3' RNA strand is identical to the coding DNA strand with the exception that thymines (T) are replaced with uracils (U) in the RNA. A coding DNA strand reading "ATG" is indirectly transcribed through the non-coding strand as "AUG" in RNA.

In prokaryotes, transcription is carried out by a single type of RNA polymerase, which needs a DNA sequence called a Pribnow box as well as a sigma factor (σ factor) to start transcription. In eukaryotes, transcription is performed by three types of RNA polymerases, each of which needs a special DNA sequence called the promoter and a set of DNA-binding proteins—transcription factors—to initiate the process. RNA polymerase I is responsible for transcription of ribosomal RNA (rRNA) genes. RNA polymerase II (Pol II) transcribes all protein-coding genes but also some non-coding RNAs (e.g., snRNAs, snoRNAs or long non-coding RNAs). Pol II includes a C-terminal domain (CTD) that is rich in serine residues. When these residues are phosphorylated, the CTD binds to various protein factors that promote transcript maturation and modification. RNA polymerase III transcribes 5S rRNA, transfer RNA (tRNA) genes, and some small non-coding RNAs (e.g., 7SK). Transcription ends when the polymerase encounters a sequence called the terminator.

RNA Processing

While transcription of prokaryotic protein-coding genes creates messenger RNA (mRNA) that is ready for translation into protein, transcription of eukaryotic genes leaves a primary transcript of RNA (pre-mRNA), which first has to undergo a series of modifications to become a mature mRNA.

These include 5' *capping*, which is set of enzymatic reactions that add 7-methylguanosine (m^7G) to the 5' end of pre-mRNA and thus protect the RNA from degradation by exonucleases. The m^7G cap is then bound by cap binding complex heterodimer (CBC20/CBC80), which aids in mRNA export to cytoplasm and also protect the RNA from decapping.

Another modification is 3' *cleavage and polyadenylation*. They occur if polyadenylation signal sequence (5'- AAUAAA-3') is present in pre-mRNA, which is usually between protein-coding sequence and terminator. The pre-mRNA is first cleaved and then a series of ~200 adenines (A) are added to form poly(A) tail, which protects the RNA from degradation. Poly(A) tail is bound by multiple poly(A)-binding proteins (PABP) necessary for mRNA export and translation re-initiation.

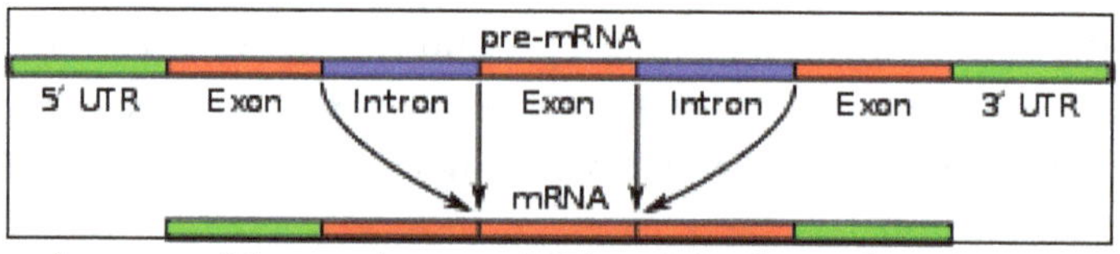

Simple illustration of exons and introns in pre-mRNA and the formation of mature mRNA by splicing. The UTRs are non-coding parts of exons at the ends of the mRNA.

A very important modification of eukaryotic pre-mRNA is *RNA splicing*. The majority of eukaryotic pre-mRNAs consist of alternating segments called exons and introns. During the process of splicing, an RNA-protein catalytical complex known as spliceosome catalyzes two transesterification reactions, which remove an intron and release it in form of lariat structure, and then splice neighbouring exons together. In certain cases, some introns or exons can be either removed or retained in mature mRNA. This so-called alternative splicing creates series of different transcripts originating from a single gene. Because these transcripts can be potentially translated into different proteins, splicing extends the complexity of eukaryotic gene expression.

Extensive RNA processing may be an evolutionary advantage made possible by the nucleus of eukaryotes. In prokaryotes, transcription and translation happen together, whilst in eukaryotes, the nuclear membrane separates the two processes, giving time for RNA processing to occur.

Non-coding RNA Maturation

In most organisms non-coding genes (ncRNA) are transcribed as precursors that undergo further processing. In the case of ribosomal RNAs (rRNA), they are often transcribed as a pre-rRNA that contains one or more rRNAs. The pre-rRNA is cleaved and modified (2′-O-methylation and pseudouridine formation) at specific sites by approximately 150 different small nucleolus-restricted RNA species, called snoRNAs. SnoRNAs associate with proteins, forming snoRNPs. While snoRNA part basepair with the target RNA and thus position the modification at a precise site, the protein part performs the catalytical reaction. In eukaryotes, in particular a snoRNP called RNase, MRP cleaves the 45S pre-rRNA into the 28S, 5.8S, and 18S rRNAs. The rRNA and RNA processing factors form large aggregates called the nucleolus.

In the case of transfer RNA (tRNA), for example, the 5' sequence is removed by RNase P, whereas the 3' end is removed by the tRNase Z enzyme and the non-templated 3' CCA tail is added by a nucleotidyl transferase. In the case of micro RNA (miRNA), miRNAs are first transcribed as primary transcripts or pri-miRNA with a cap and poly-A tail and processed to short, 70-nucleotide stem-loop structures known as pre-miRNA in the cell nucleus by the enzymes Drosha and Pasha. After being exported, it is then processed to mature miRNAs in the cytoplasm by interaction with the endonuclease Dicer, which also initiates the formation of the RNA-induced silencing complex (RISC), composed of the Argonaute protein.

Even snRNAs and snoRNAs themselves undergo series of modification before they become part of functional RNP complex. This is done either in the nucleoplasm or in the specialized compartments called Cajal bodies. Their bases are methylated or pseudouridinilated by a group of small Cajal body-specific RNAs (scaRNAs), which are structurally similar to snoRNAs.

RNA Export

In eukaryotes most mature RNA must be exported to the cytoplasm from the nucleus. While some RNAs function in the nucleus, many RNAs are transported through the nuclear pores and into the cytosol. Notably this includes all RNA types involved in protein synthesis. In some cases RNAs are additionally transported to a specific part of the cytoplasm, such as a synapse; they are then towed by motor proteins that bind through linker proteins to specific sequences (called "zipcodes") on the RNA.

Translation

For some RNA (non-coding RNA) the mature RNA is the final gene product. In the case of messenger RNA (mRNA) the RNA is an information carrier coding for the synthesis of one or more proteins. mRNA carrying a single protein sequence (common in eukaryotes) is monocistronic whilst mRNA carrying multiple protein sequences (common in prokaryotes) is known as polycistronic.

Every mRNA consists of three parts: a 5' untranslated region (5'UTR), a protein-coding region or open reading frame (ORF), and a 3' untranslated region (3'UTR). The coding region carries information for protein synthesis encoded by the genetic code to form triplets. Each triplet of nucleotides of the coding region is called a codon and corresponds to a binding site complementary to an anticodon triplet in transfer RNA. Transfer RNAs with the same anticodon sequence always carry an identical type of amino acid. Amino acids are then chained together by the ribosome according to the order of triplets in the coding region. The ribosome helps transfer RNA to bind to messenger RNA and takes the amino acid from each transfer RNA and makes a structure-less protein out of it. Each mRNA molecule is translated into many protein molecules, on average ~2800 in mammals.

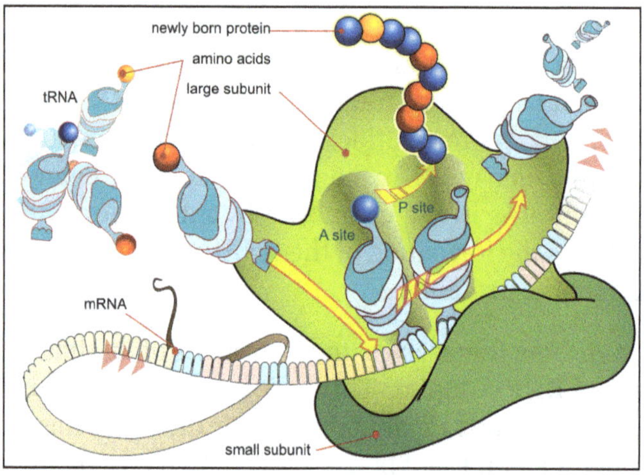

During the translation, tRNA charged with amino acid enters the ribosome and aligns with the correct mRNA triplet. Ribosome then adds amino acid to growing protein chain.

In prokaryotes translation generally occurs at the point of transcription (co-transcriptionally), often using a messenger RNA that is still in the process of being created. In eukaryotes translation can occur in a variety of regions of the cell depending on where the protein being written is supposed to be. Major locations are the cytoplasm for soluble cytoplasmic proteins and the membrane of the endoplasmic reticulum for proteins that are for export from the cell or insertion into a cell membrane. Proteins that are supposed to be expressed at the endoplasmic reticulum are recognised part-way through the translation process. This is governed by the signal recognition particle—a protein that binds to the ribosome and directs it to the endoplasmic reticulum when it finds a signal peptide on the growing (nascent) amino acid chain. Translation is the communication of the meaning of a source-language text by means of an equivalent target-language text.

Folding

The polypeptide folds into its characteristic and functional three-dimensional structure from a random coil. Each protein exists as an unfolded polypeptide or random coil when translated from a sequence of mRNA into a linear chain of amino acids. This polypeptide lacks any developed three-dimensional structure (the left hand side of the neighboring figure). Amino acids interact with each other to produce a well-defined three-dimensional structure, the folded protein (the right hand side of the figure) known as the native state. The resulting three-dimensional structure is determined by the amino acid sequence (Anfinsen's dogma).

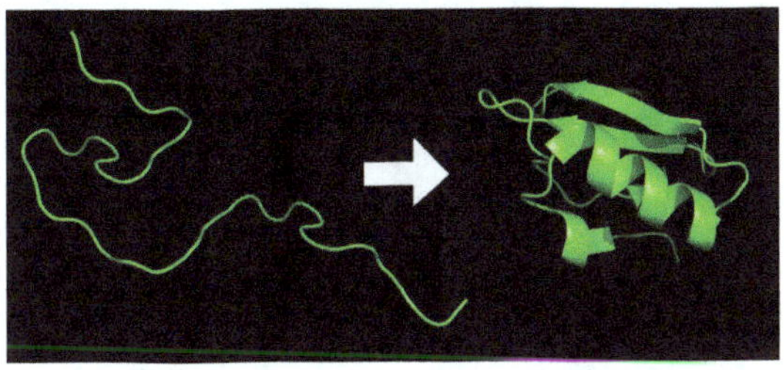

Protein before (left) and after (right) folding.

The correct three-dimensional structure is essential to function, although some parts of functional proteins may remain unfolded Failure to fold into the intended shape usually produces inactive proteins with different properties including toxic prions. Several neurodegenerative and other diseases are believed to result from the accumulation of *misfolded* proteins. Many allergies are caused by the folding of the proteins, for the immune system does not produce antibodies for certain protein structures.

Enzymes called chaperones assist the newly formed protein to attain (fold into) the 3-dimensional structure it needs to function. Similarly, RNA chaperones help RNAs attain their functional shapes. Assisting protein folding is one of the main roles of the endoplasmic reticulum in eukaryotes.

Translocation

Secretory proteins of eukaryotes or prokaryotes must be translocated to enter the secretory pathway. Newly synthesized proteins are directed to the eukaryotic Sec61 or prokaryotic SecYEG translocation channel by signal peptides. The efficiency of protein secretion in eukaryotes is very dependent on the signal peptide which has been used.

Protein transport

Many proteins are destined for other parts of the cell than the cytosol and a wide range of signalling sequences or (signal peptides) are used to direct proteins to where they are supposed to be. In prokaryotes this is normally a simple process due to limited compartmentalisation of the cell. However, in eukaryotes there is a great variety of different targeting processes to ensure the protein arrives at the correct organelle.

Not all proteins remain within the cell and many are exported, for example, digestive enzymes, hormones and extracellular matrix proteins. In eukaryotes the export pathway is well developed and the main mechanism for the export of these proteins is translocation to the endoplasmic reticulum, followed by transport via the Golgi apparatus.

Regulation of Gene Expression

The patchy colours of a tortoiseshell cat are the result of different levels of expression of pigmentation genes in different areas of the skin.

Regulation of gene expression refers to the control of the amount and timing of appearance of the functional product of a gene. Control of expression is vital to allow a cell to produce the gene products it needs when it needs them; in turn, this gives cells the flexibility to adapt to a variable environment, external signals, damage to the cell, and other stimuli. More generally, gene regulation gives the cell control over all structure and function, and is the basis for cellular differentiation, morphogenesis and the versatility and adaptability of any organism.

Numerous terms are used to describe types of genes depending on how they are regulated; these include:

- A constitutive gene is a gene that is transcribed continually as opposed to a facultative gene, which is only transcribed when needed.
- A *housekeeping gene* is a gene that is required to maintain basic cellular function and so is typically expressed in all cell types of an organism. Examples include actin, GAPDH and ubiquitin. Some housekeeping genes are transcribed at a relatively constant rate and these genes can be used as a reference point in experiments to measure the expression rates of other genes.
- A facultative gene is a gene only transcribed when needed as opposed to a constitutive gene.
- An inducible gene is a gene whose expression is either responsive to environmental change or dependent on the position in the cell cycle.

Any step of gene expression may be modulated, from the DNA-RNA transcription step to post-translational modification of a protein. The stability of the final gene product, whether it is RNA or protein, also contributes to the expression level of the gene—an unstable product results in a low expression level. In general gene expression is regulated through changes in the number and type of interactions between molecules that collectively influence transcription of DNA and translation of RNA.

Some simple examples of where gene expression is important are:

- Control of insulin expression so it gives a signal for blood glucose regulation.
- X chromosome inactivation in female mammals to prevent an "overdose" of the genes it contains.
- Cyclin expression levels control progression through the eukaryotic cell cycle.

Transcriptional Regulation

Regulation of transcription can be broken down into three main routes of influence; genetic (direct interaction of a control factor with the gene), modulation interaction of a control factor with the transcription machinery and epigenetic (non-sequence changes in DNA structure that influence transcription).

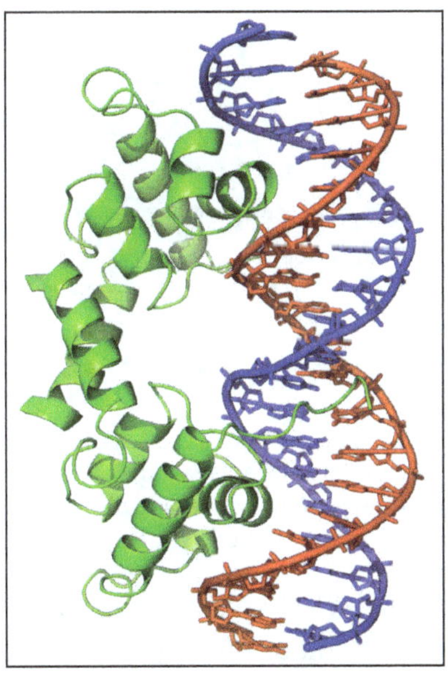

The lambda repressor transcription factor (green) binds as a dimer to major groove of DNA target (red and blue) and disables initiation of transcription. From PDB: 1LMB.

Direct interaction with DNA is the simplest and the most direct method by which a protein changes transcription levels. Genes often have several protein binding sites around the coding region with the specific function of regulating transcription. There are many classes of regulatory DNA binding sites known as enhancers, insulators and silencers. The mechanisms for regulating transcription are very varied, from blocking key binding sites on the DNA for RNA polymerase to acting as an activator and promoting transcription by assisting RNA polymerase binding.

The activity of transcription factors is further modulated by intracellular signals causing protein post-translational modification including phosphorylated, acetylated, or glycosylated. These changes influence a transcription factor's ability to bind, directly or indirectly, to promoter DNA, to recruit RNA polymerase, or to favor elongation of a newly synthesized RNA molecule.

The nuclear membrane in eukaryotes allows further regulation of transcription factors by the duration of their presence in the nucleus, which is regulated by reversible changes in their structure and by binding of other proteins. Environmental stimuli or endocrine signals may cause modification of regulatory proteins eliciting cascades of intracellular signals, which result in regulation of gene expression.

More recently it has become apparent that there is a significant influence of non-DNA-sequence specific effects on transcription. These effects are referred to as epigenetic and involve the higher order structure of DNA, non-sequence specific DNA binding proteins and chemical modification of DNA. In general epigenetic effects alter the accessibility of DNA to proteins and so modulate transcription.

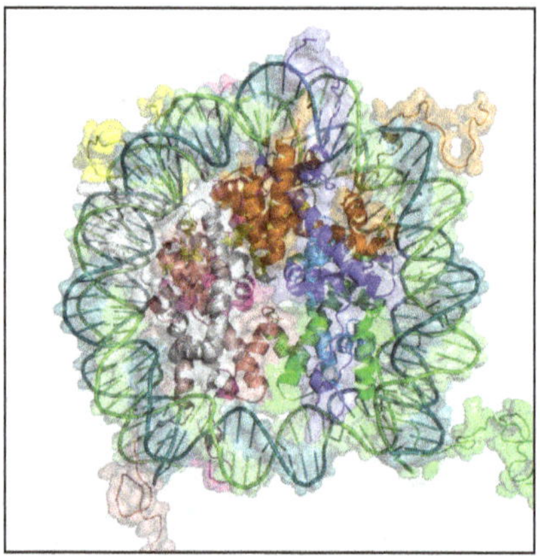

In eukaryotes, DNA is organized in form of nucleosomes. Note how the DNA (blue and green) is tightly wrapped around the protein core made of histone octamer (ribbon coils), restricting access to the DNA. From PDB: 1KX5.

DNA methylation is a widespread mechanism for epigenetic influence on gene expression and is seen in bacteria and eukaryotes and has roles in heritable transcription silencing and transcription regulation. In eukaryotes the structure of chromatin, controlled by the histone code, regulates access to DNA with significant impacts on the expression of genes in euchromatin and heterochromatin areas.

Transcriptional Regulation in Cancer

The majority of gene promoters contain a CpG island with numerous CpG sites. When many of a gene's promoter CpG sites are methylated the gene becomes silenced. Colorectal cancers typically have 3 to 6 driver mutations and 33 to 66 hitchhiker or passenger mutations. However, transcriptional silencing may be of more importance than mutation in causing progression to cancer. For example, in colorectal cancers about 600 to 800 genes are transcriptionally silenced by CpG island methylation. Transcriptional repression in cancer can also occur by other epigenetic mechanisms, such as altered expression of microRNAs. In breast cancer, transcriptional repression of BRCA1 may occur more frequently by over-expressed microRNA-182 than by hypermethylation of the BRCA1 promoter.

Post-transcriptional Regulation

In eukaryotes, where export of RNA is required before translation is possible, nuclear export is thought to provide additional control over gene expression. All transport in and out of the nucleus is via the nuclear pore and transport is controlled by a wide range of importin and exportin proteins.

Expression of a gene coding for a protein is only possible if the messenger RNA carrying the code survives long enough to be translated. In a typical cell, an RNA molecule is only stable if specifically protected from degradation. RNA degradation has particular importance in regulation of expression in eukaryotic cells where mRNA has to travel significant distances before being translated. In eukaryotes, RNA is stabilised by certain post-transcriptional modifications, particularly the 5' cap and poly-adenylated tail.

Intentional degradation of mRNA is used not just as a defence mechanism from foreign RNA (normally from viruses) but also as a route of mRNA *destabilisation*. If an mRNA molecule has a complementary sequence to a small interfering RNA then it is targeted for destruction via the RNA interference pathway.

Three Prime Untranslated Regions and MicroRNAs

Three prime untranslated regions (3'UTRs) of messenger RNAs (mRNAs) often contain regulatory sequences that post-transcriptionally influence gene expression. Such 3'-UTRs often contain both binding sites for microRNAs (miRNAs) as well as for regulatory proteins. By binding to specific sites within the 3'-UTR, miRNAs can decrease gene expression of various mRNAs by either inhibiting translation or directly causing degradation of the transcript. The 3'-UTR also may have silencer regions that bind repressor proteins that inhibit the expression of a mRNA.

The 3'-UTR often contains microRNA response elements (MREs). MREs are sequences to which miRNAs bind. These are prevalent motifs within 3'-UTRs. Among all regulatory motifs within the 3'-UTRs (e.g. including silencer regions), MREs make up about half of the motifs.

As of 2014, the miRBase web site, an archive of miRNA sequences and annotations, listed 28,645 entries in 233 biologic species. Of these, 1,881 miRNAs were in annotated human miRNA loci. miRNAs were predicted to have an average of about four hundred target mRNAs (affecting expression of several hundred genes). Freidman et al.

estimate that >45,000 miRNA target sites within human mRNA 3'UTRs are conserved above background levels, and >60% of human protein-coding genes have been under selective pressure to maintain pairing to miRNAs.

Direct experiments show that a single miRNA can reduce the stability of hundreds of unique mRNAs. Other experiments show that a single miRNA may repress the production of hundreds of proteins, but that this repression often is relatively mild (less than 2-fold).

The effects of miRNA dysregulation of gene expression seem to be important in cancer. For instance, in gastrointestinal cancers, nine miRNAs have been identified as epigenetically altered and effective in down regulating DNA repair enzymes.

The effects of miRNA dysregulation of gene expression also seem to be important in neuropsychiatric disorders, such as schizophrenia, bipolar disorder, major depression, Parkinson's disease, Alzheimer's disease and autism spectrum disorders.

Translational Regulation

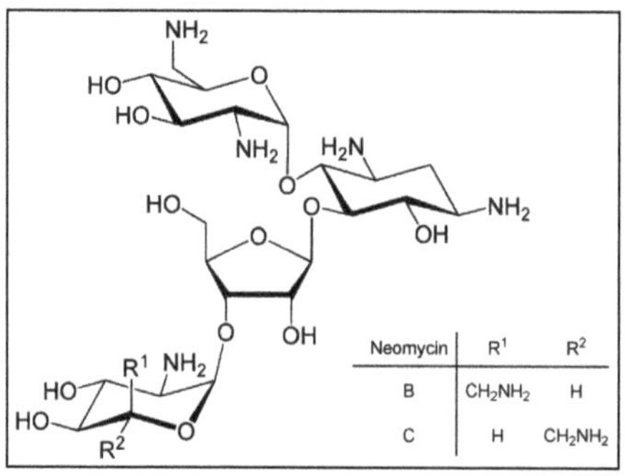

Neomycin is an example of a small molecule that reduces expression of all protein genes inevitably leading to cell death; it thus acts as an antibiotic.

Direct regulation of translation is less prevalent than control of transcription or mRNA stability but is occasionally used. Inhibition of protein translation is a major target for toxins and antibiotics, so they can kill a cell by overriding its normal gene expression control. Protein synthesis inhibitors include the antibiotic neomycin and the toxin ricin.

Protein Degradation

Once protein synthesis is complete, the level of expression of that protein can be reduced by protein degradation. There are major protein degradation pathways in all prokaryotes and eukaryotes, of which the proteasome is a common component. An unneeded or damaged protein is often labeled for degradation by addition of ubiquitin.

Measurement

Measuring gene expression is an important part of many life sciences, as the ability to quantify the level at which a particular gene is expressed within a cell, tissue or organism can provide a lot of valuable information. For example, measuring gene expression can:

- Identify viral infection of a cell (viral protein expression).
- Determine an individual's susceptibility to cancer (oncogene expression).
- Find if a bacterium is resistant to penicillin (beta-lactamase expression).

Similarly, the analysis of the location of protein expression is a powerful tool, and this can be done on an organismal or cellular scale. Investigation of localization is particularly important for the study of development in multicellular organisms and as an indicator of protein function in single cells. Ideally, measurement of expression is done by detecting the final gene product (for many genes, this is the protein); however, it is often easier to detect one of the precursors, typically mRNA and to infer gene-expression levels from these measurements.

mRNA quantification

Levels of mRNA can be quantitatively measured by northern blotting, which provides size and sequence information about the mRNA molecules. A sample of RNA is separated on an agarose gel and hybridized to a radioactively labeled RNA probe that is complementary to the target sequence. The radiolabeled RNA is then detected by an autoradiograph. Because the use of radioactive reagents makes the procedure time consuming and potentially dangerous, alternative labeling and detection methods, such as digoxigenin and biotin chemistries, have been developed. Perceived disadvantages of Northern blotting are that large quantities of RNA are required and that quantification may not be completely accurate, as it involves measuring band strength in an image of a gel. On the other hand, the additional mRNA size information from the Northern blot allows the discrimination of alternately spliced transcripts.

Another approach for measuring mRNA abundance is RT-qPCR. In this technique, reverse transcription is followed by quantitative PCR. Reverse transcription first generates a DNA template from the mRNA; this single-stranded template is called cDNA. The cDNA template is then amplified in the quantitative step, during which the fluorescence emitted by labeled hybridization probes or intercalating dyes changes as the DNA amplification process progresses. With a carefully constructed standard curve, qPCR can produce an absolute measurement of the number of copies of original mRNA, typically in units of copies per nanolitre of homogenized tissue or copies per cell. qPCR is very sensitive (detection of a single mRNA molecule is theoretically possible), but can be expensive depending on the type of reporter used; fluorescently labeled oligonucleotide probes are more expensive than non-specific intercalating fluorescent dyes.

For expression profiling, or high-throughput analysis of many genes within a sample, quantitative PCR may be performed for hundreds of genes simultaneously in the case of low-density arrays. A second approach is the hybridization microarray. A single array or "chip" may contain probes to determine transcript levels for every known gene in the genome of one or more organisms. Alternatively, "tag based" technologies like Serial analysis of gene expression (SAGE) and RNA-Seq, which can provide a relative measure of the cellular concentration of different mRNAs, can be used. An advantage of tag-based methods is the "open architecture", allowing for the exact measurement of any transcript, with a known or unknown sequence. Next-generation sequencing (NGS) such as RNA-Seq is another approach, producing vast quantities of sequence data that can be matched to a reference genome. Although NGS is comparatively time-consuming, expensive, and resource-intensive, it can identify single-nucleotide polymorphisms, splice-variants, and novel genes, and can also be used to profile expression in organisms for which little or no sequence information is available.

Protein Quantification

For genes encoding proteins, the expression level can be directly assessed by a number of methods with some clear analogies to the techniques for mRNA quantification.

The most commonly used method is to perform a Western blot against the protein of interest—this gives information on the size of the protein in addition to its identity. A sample (often cellular lysate) is separated on a polyacrylamide gel, transferred to a membrane and then probed with an antibody to the protein of interest. The antibody can either be conjugated to a fluorophore or to horseradish peroxidase for imaging and/or quantification. The gel-based nature of this assay makes quantification less accurate, but it has the advantage of being able to identify later modifications to the protein, for example proteolysis or ubiquitination, from changes in size.

Localisation

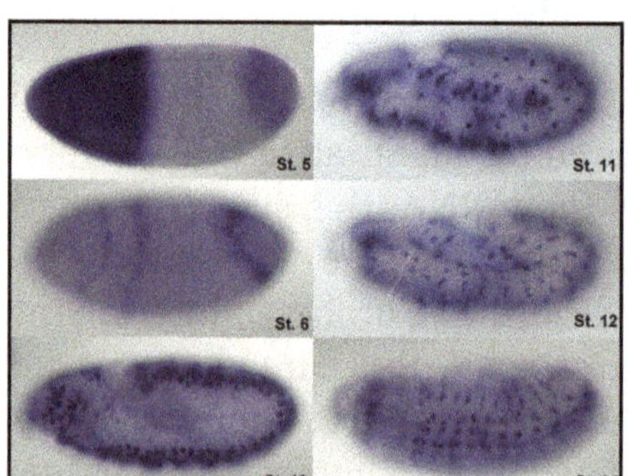

In situ-hybridization of Drosophila embryos at different developmental stages for the mRNA responsible for the expression of hunchback. High intensity of blue color marks places with high hunchback mRNA quantity.

Analysis of expression is not limited to quantification; localisation can also be determined. mRNA can be detected with a suitably labelled complementary mRNA strand

and protein can be detected via labelled antibodies. The probed sample is then observed by microscopy to identify where the mRNA or protein is.

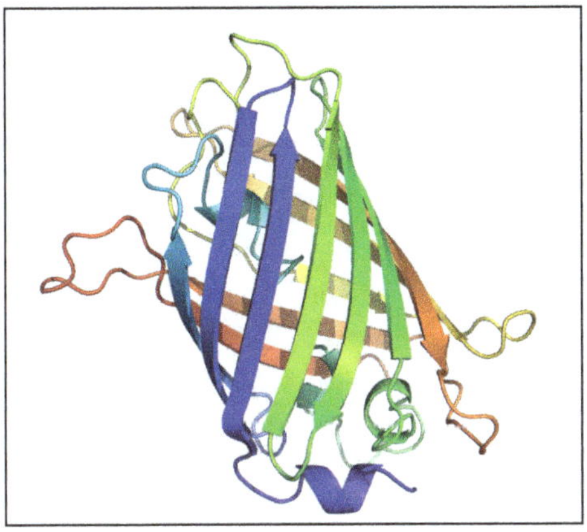

The three-dimensional structure of green fluorescent protein. The residues in the centre of the "barrel" are responsible for production of green light after exposing to higher energetic blue light. From PDB: 1EMA.

By replacing the gene with a new version fused to a green fluorescent protein (or similar) marker, expression may be directly quantified in live cells. This is done by imaging using a fluorescence microscope. It is very difficult to clone a GFP-fused protein into its native location in the genome without affecting expression levels so this method often cannot be used to measure endogenous gene expression. It is, however, widely used to measure the expression of a gene artificially introduced into the cell, for example via an expression vector. It is important to note that by fusing a target protein to a fluorescent reporter the protein's behavior, including its cellular localization and expression level, can be significantly changed.

The enzyme-linked immunosorbent assay works by using antibodies immobilised on a microtiter plate to capture proteins of interest from samples added to the well. Using a detection antibody conjugated to an enzyme or fluorophore the quantity of bound protein can be accurately measured by fluorometric or colourimetric detection. The detection process is very similar to that of a Western blot, but by avoiding the gel steps more accurate quantification can be achieved.

Expression System

An expression system is a system specifically designed for the production of a gene product of choice. This is normally a protein although may also be RNA, such as tRNA or a ribozyme. An expression system consists of a gene, normally encoded by DNA, and the molecular machinery required to transcribe the DNA into mRNA and translate the mRNA into protein using the reagents provided. In the broadest sense this includes every living cell but the term is more normally used to refer to expression as a laboratory tool. An expression system is therefore often artificial in some manner. Expression systems are, however, a fundamentally natural process. Viruses are an excellent example where they replicate by using the host cell as an expression system for the viral proteins and genome.

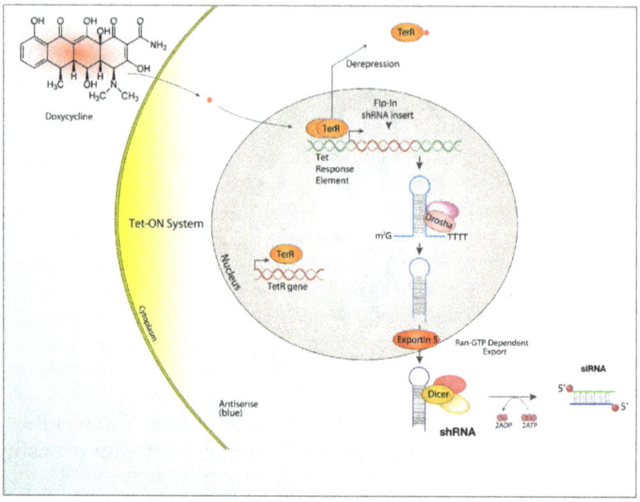

Tet-ON inducible shRNA system.

Inducible Expression

Doxycycline is also used in "Tet-on" and "Tet-off" tetracycline controlled transcriptional activation to regulate transgene expression in organisms and cell cultures.

In Nature

In addition to these biological tools, certain naturally observed configurations of DNA (genes, promoters, enhancers, repressors) and the associated machinery itself are referred to as an expression system. This term is normally used in the case where a gene or set of genes is switched on under well defined conditions, for example, the simple repressor switch expression system in Lambda phage and the lac operator system in bacteria. Several natural expression systems are directly used or modified and used for artificial expression systems such as the Tet-on and Tet-off expression system.

Gene Networks

Genes have sometimes been regarded as nodes in a network, with inputs being proteins such as transcription factors, and outputs being the level of gene expression. The node itself performs a function, and the operation of these functions have been interpreted as performing a kind of information processing within cells and determines cellular behavior.

Gene networks can also be constructed without formulating an explicit causal model. This is often the case when assembling networks from large expression data sets. Covariation and correlation of expression is computed across a large sample of cases and measurements (often transcriptome or proteome data). The source of variation can be either experimental or natural (observational). There are several ways to construct gene expression networks, but one common approach is to compute a matrix of all pair-wise correlations of expression across conditions, time points, or individuals and convert the matrix (after thresholding at some cut-off value) into a graphical representation in which nodes represent genes, transcripts, or proteins and edges connecting these nodes represent the strength of association. Weighted correlation network analysis involves weighted networks defined by soft-thresholding the pairwise correlations among variables (e.g. measures of transcript abundance).

Techniques and Tools

The following experimental techniques are used to measure gene expression and are listed in roughly chronological order, starting with the older, more established technologies. They are divided into two groups based on their degree of multiplexity.

- Low-to-mid-plex techniques:
    - Reporter gene.
    - Northern blot.
    - Western blot.
    - Fluorescent in situ hybridization.
    - Reverse transcription PCR.
- Higher-plex techniques:
    - SAGE.
    - DNA microarray.
    - Tiling array.
    - RNA-Seq.

Gene Regulatory Network

A gene (or genetic) regulatory network (GRN) is a collection of molecular regulators that interact with each other and with other substances in the cell to govern the gene expression levels of mRNA and proteins. These play a central role in morphogenesis, the creation of body structures, which in turn is central to evolutionary developmental biology (evo-devo).

The regulator can be DNA, RNA, protein and complexes of these. The interaction can be direct or indirect (through transcribed RNA or translated protein). In general, each mRNA molecule goes on to make a specific protein (or set of proteins). In some cases this protein will be structural, and will accumulate at the cell membrane or within the cell to give it particular structural properties. In other cases the protein will be an enzyme, i.e., a micro-machine that catalyses a certain reaction, such as the breakdown of a food source or toxin. Some proteins though serve only to activate other genes, and these are the transcription factors that are the main players in regulatory networks or cascades. By binding to the promoter region at the start of other genes they turn them on, initiating the production of another protein, and so on. Some transcription factors are inhibitory.

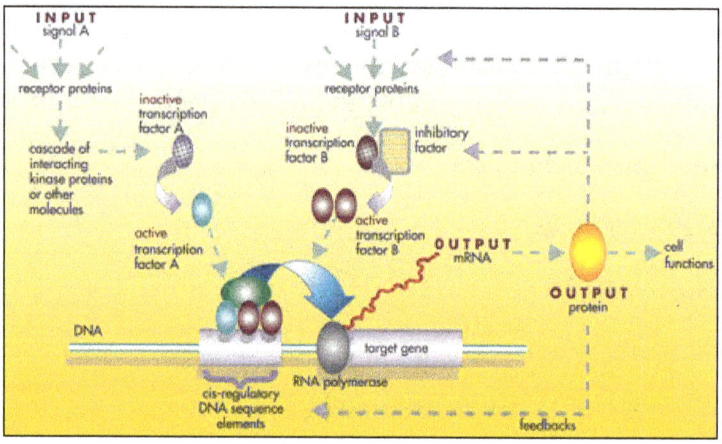

Structure of a gene regulatory network.

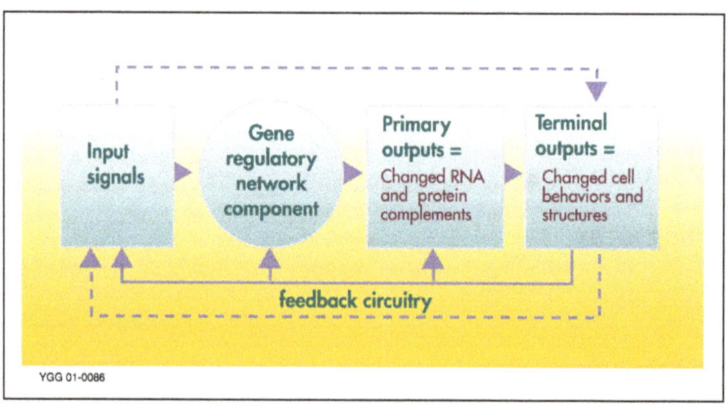

Control process of a gene regulatory network.

In single-celled organisms, regulatory networks respond to the external environment, optimising the cell at a given time for survival in this environment. Thus a yeast cell, finding itself in a sugar solution, will turn on genes to make enzymes that process the sugar to alcohol. This process, which we associate with wine-making, is how the yeast cell makes its living, gaining energy to multiply, which under normal circumstances would enhance its survival prospects.

In multicellular animals the same principle has been put in the service of gene cascades that control body-shape. Each time a cell divides, two cells result which, although they contain the same genome in full, can differ in which genes are turned on and making proteins. Sometimes a 'self-sustaining feedback loop' ensures that a cell maintains its identity and passes it on. Less understood is the mechanism of epigenetics by which chromatin modification may provide cellular memory by blocking or allowing transcription. A major feature of multicellular animals is the use of morphogen gradients, which in effect provide a positioning system that tells a cell where in the body it is, and hence what sort of cell to become. A gene that is turned on in one cell may make a product that leaves the cell and diffuses through adjacent cells, entering them and turning on genes only when it is present above a certain threshold level. These cells are thus induced into a new fate, and may even generate other morphogens that signal back to the original cell. Over longer distances morphogens may use the active process of signal transduction. Such signalling controls embryogenesis, the building of a body plan from scratch through a series of sequential steps. They also control and maintain adult bodies through feedback processes, and the loss of such feedback because of a mutation can be responsible for the cell proliferation that is seen in cancer. In parallel with this process of building structure, the gene cascade turns on genes that make structural proteins that give each cell the physical properties it needs.

Overview

At one level, biological cells can be thought of as "partially mixed bags" of biological chemicals – in the discussion of gene regulatory networks, these chemicals are mostly the messenger RNAs (mRNAs) and proteins that arise from gene expression. These mRNA and proteins interact with each other with various degrees of specificity. Some diffuse around the cell. Others are bound to cell membranes, interacting with molecules in the environment. Still others pass through cell membranes and mediate long range signals to other cells in a multi-cellular organism. These molecules and their interactions comprise a *gene regulatory network*. A typical gene regulatory network looks something like this:

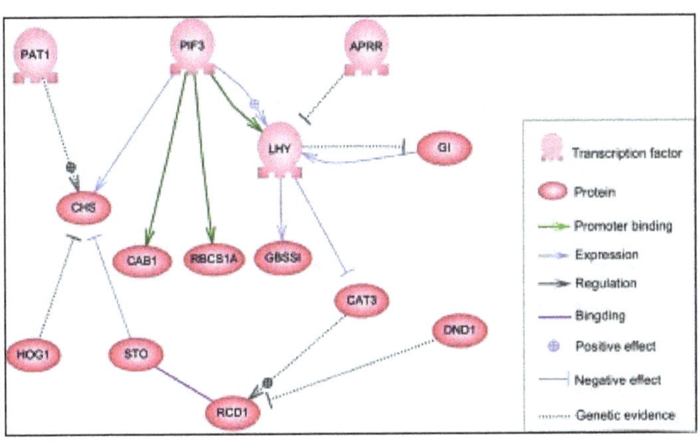

Example of a regulatory network.

The nodes of this network are proteins, their corresponding mRNAs, and protein/protein complexes. Nodes that are depicted as lying along vertical lines are associated with the cell/environment interfaces, while the others are free-floating and can diffuse. Implied are genes, the DNA sequences which are transcribed into the mRNAs that translate into proteins. Edges between nodes represent individual molecular reactions, the protein/protein and protein/mRNA interactions through which the products of one gene affect those of another, though the lack of experimentally obtained information often implies that some reactions are not modeled at such a fine level of detail. These interactions can be inductive (the arrowheads), with an increase in the concentration of one leading to an increase in the other, or inhibitory (the filled circles), with an increase in one leading to a decrease in the other. A series of edges indicates a chain of such dependencies, with cycles corresponding to feedback loops. The network structure is an abstraction of the system's chemical dynamics, describing the manifold ways in which one substance affects all the others to which it is connected. In practice, such GRNs are inferred from the biological literature on a given system and represent a distillation of the collective knowledge about a set of related biochemical reactions. To speed up the manual curation of GRNs, some recent efforts try to use text mining and information extraction technologies for this purpose.

Genes can be viewed as nodes in the network, with input being proteins such as transcription factors, and outputs being the level of gene expression. The node itself can also be viewed as a function which can be obtained by combining basic functions upon the inputs (in the Boolean network described below these are Boolean functions, typically AND, OR, and NOT). These functions have been interpreted as performing a kind of information processing within the cell, which determines cellular behavior. The basic drivers within cells are concentrations of some proteins, which determine both spatial (location within the cell or tissue) and temporal (cell cycle or developmental stage) coordinates of the cell, as a kind of "cellular memory". The gene networks are only beginning to be understood, and it is a next step for biology to attempt to deduce the functions for each gene "node", to help understand the behavior of the system in increasing levels of complexity, from gene to signaling pathway, cell or tissue level.

Mathematical models of GRNs have been developed to capture the behavior of the system being modeled, and in some cases generate predictions corresponding with experimental observations. In some other cases, models have proven to make accurate novel predictions, which can be tested experimentally, thus suggesting new approaches to explore in an experiment that sometimes wouldn't be considered in the design of the protocol of an experimental laboratory. The most common modeling technique involves the use of coupled ordinary differential equations (ODEs). Several other promising modeling techniques have been used, including Boolean networks, Petri nets, Bayesian networks, graphical Gaussian models, Stochastic, and Process Calculi. Conversely, techniques have been proposed for generating models of GRNs that best explain a set of time series observations. Recently it has been shown that ChIP-seq signal of Histone modification are more correlated with transcription factor motifs at promoters in comparison to RNA level. Hence it is proposed that time-series histone modification ChIP-seq could provide more reliable inference of gene-regulatory networks in comparison to methods based on expression levels.

Structure and Evolution

Global Feature

Gene regulatory networks are generally thought to be made up of a few highly connected nodes (hubs) and many poorly connected nodes nested within a hierarchical regulatory regime. Thus gene regulatory networks approximate a hierarchical scale free network topology. This is consistent with the view that most genes have limited pleiotropy and operate within regulatory modules. This structure is thought to evolve due to the preferential attachment of duplicated genes to more highly connected genes. Recent work has also shown that natural selection tends to favor networks with sparse connectivity.

There are primarily two ways that networks can evolve, both of which can occur simultaneously. The first is that network topology can be changed by the addition or subtraction of nodes (genes) or parts of the network (modules) may be expressed in different contexts. The *Drosophila* Hippo signaling pathway provides a good example. The Hippo signaling pathway controls both mitotic growth and post-mitotic cellular differentiation. Recently it was found that the network the Hippo signaling pathway operates in differs between these two functions which in turn changes the behavior of the Hippo signaling pathway. This suggests that the Hippo signaling pathway operates as a conserved regulatory module that can be used for multiple functions depending on context. Thus, changing network topology can allow a conserved module to serve multiple functions and alter the final output of the network. The second way networks can evolve is by changing the strength of interactions between nodes, such as how strongly a transcription factor may bind to a cis-regulatory element. Such variation in strength of network edges has been shown to underlie between species variation in vulva cell fate patterning of *Caenorhabditis* worms.

Local Feature

Another widely cited characteristic of gene regulatory network is their abundance of certain repetitive sub-networks known as network motifs. Network motifs can be regarded as repetitive topological patterns when dividing a big network into small blocks. Previous analysis found several types of motifs that appeared more often in gene regulatory networks than in randomly generated networks. As an example, one such motif is called feed-forward loops, which consist three nodes. This motif is the most abundant among all possible motifs made up of three nodes, as is shown in the gene regulatory networks of fly, nematode, and human.

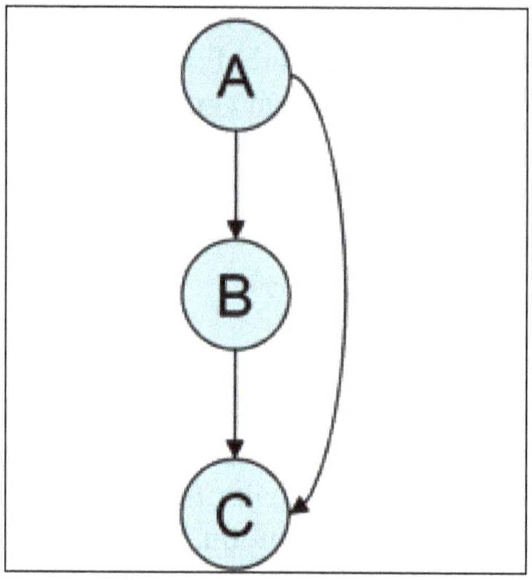

Feed-forward loop.

The enriched motifs have been proposed to follow convergent evolution, suggesting they are "optimal designs" for certain regulatory purposes. For example, modeling shows that feed-forward loops are able to coordinate the change in node A (in terms of concentration and activity) and the expression dynamics of node C, creating different input-output behaviors. The galactose utilization system of *E. coli* contains a feed-forward loop which accelerates the activation of galactose utilization operon *galETK*, potentially facilitating the metabolic transition to galactose when glucose is depleted. The feed-forward loop in the arabinose utilization systems of *E.coli* delays the activation of arabinose catabolism operon and transporters, potentially avoiding unnecessary metabolic transition due to temporary fluctuations in upstream signaling pathways. Similarly in the Wnt signaling pathway of *Xenopus*, the feed-forward loop acts as a fold-change detector that responses to the fold change, rather than the absolute change, in the level of β-catenin, potentially increasing the resistance to fluctuations in β-catenin levels. Following the convergent evolution hypothesis, the enrichment of feed-forward loops would be an adaptation for fast response and noise resistance. A recent research found that yeast grown in an environment of constant glucose developed mutations in glucose signaling pathways and growth regulation pathway, suggesting regulatory components responding to environmental changes are dispensable under constant environment.

On the other hand, some researchers hypothesize that the enrichment of network motifs is non-adaptive. In other words, gene regulatory networks can evolve to a similar structure without the specific selection on the proposed input-output behavior. Support for this hypothesis often comes from computational simulations. For example, fluctuations in the abundance of feed-forward loops in a model that simulates the evolution of gene regulatory networks by randomly rewiring nodes may suggest that the enrichment of feed-forward loops is a side-effect of evolution. In another model of gene regulator networks evolution, the ratio of the frequencies of gene duplication and gene deletion show great influence on network topology: certain ratios lead to the enrichment of feed-forward loops and create networks that show features of hierarchical scale free networks.

Bacterial Regulatory Networks

Regulatory networks allow bacteria to adapt to almost every environmental niche on earth. A network of interactions among diverse types of molecules including DNA, RNA, proteins and metabolites, is utilised by the bacteria to achieve regulation of gene expression. In bacteria, the principal function of regulatory networks is to control the response to environmental changes, for example nutritional status and environmental stress. A complex organization of networks permits the microorganism to coordinate and integrate multiple environmental signals.

Modelling

Coupled Ordinary Differential Equations

It is common to model such a network with a set of coupled ordinary differential equations (ODEs) or SDEs, describing the reaction kinetics of the constituent parts. Suppose that our regulatory network has N nodes, and let $S_1(t), S_2(t), \ldots, S_N(t)$ represent the concentrations of the N corresponding substances at time t. Then the temporal evolution of the system can be described approximately by:

$$\frac{dS_j}{dt} = f_j\left(S_1, S_2, \ldots, S_N\right)$$

where the functions f_j express the dependence of S_j on the concentrations of other substances present in the cell. The functions f_j are ultimately derived from basic principles of chemical kinetics or simple expressions derived from these e.g. Michaelis-Menten enzymatic kinetics. Hence, the functional forms of the f_j are usually chosen as low-order polynomials or Hill functions that serve as an ansatz for the real molecular dynamics. Such models are then studied using the mathematics of nonlinear dynamics. System-specific information, like reaction rate constants and sensitivities, are encoded as constant parameters.

By solving for the fixed point of the system:

$$\frac{dS_j}{dt} = 0$$

for all j, one obtains (possibly several) concentration profiles of proteins and mRNAs that are theoretically sustainable (though not necessarily stable). Steady states of kinetic equations thus correspond to potential cell types, and oscillatory solutions to the above equation to naturally cyclic cell types. Mathematical stability of these attractors can usually be characterized by the sign of higher derivatives at critical points, and then correspond to biochemical stability of the concentration profile. Critical points and bifurcations in the equations correspond to critical cell states in which small state or parameter perturbations could switch the system between one of several stable differentiation fates. Trajectories correspond to the unfolding of biological pathways and transients of the equations to short-term biological events.

Boolean Network

The following example illustrates how a Boolean network can model a GRN together with its gene products (the outputs) and the substances from the environment that affect it (the inputs). Stuart Kauffman was amongst the first biologists to use the metaphor of Boolean networks to model genetic regulatory networks.

1. Each gene, each input, and each output is represented by a node in a directed graph in which there is an arrow from one node to another if and only if there is a causal link between the two nodes.
2. Each node in the graph can be in one of two states: on or off.
3. For a gene, "on" corresponds to the gene being expressed; for inputs and outputs, "off" corresponds to the substance being present.
4. Time is viewed as proceeding in discrete steps. At each step, the new state of a node is a Boolean function of the prior states of the nodes with arrows pointing towards it.

The validity of the model can be tested by comparing simulation results with time series observations. A partial validation of a Boolean network model can also come from testing the predicted existence of a yet unknown regulatory connection between two particular transcription factors that each are nodes of the model.

Continuous Networks

Continuous network models of GRNs are an extension of the boolean networks described above. Nodes still represent genes and connections between them regulatory influences on gene expression. Genes in biological systems display a continuous range of activity levels and it has been argued that using a continuous representation captures several properties of gene regulatory networks not present in the Boolean model. Formally most of these approaches are similar to an artificial neural network, as inputs to a node are summed up and the result serves as input to a sigmoid function, e.g., but proteins do often control gene expression in a synergistic, i.e. non-linear, way. However, there is now a continuous network model that allows grouping of inputs to a node thus realizing another level of regulation. This model is formally closer to a higher order recurrent neural network. The same model has also been used to mimic the evolution of cellular differentiation and even multicellular morphogenesis.

Stochastic Gene Networks

Recent experimental results have demonstrated that gene expression is a stochastic process. Thus, many authors are now using the stochastic formalism, after the work by. Works on single gene expression and small synthetic genetic networks, such as the genetic toggle switch of Tim Gardner and Jim Collins, provided additional experimental data on the phenotypic variability and the stochastic nature of gene expression. The first versions of stochastic models of gene expression involved only instantaneous reactions and were driven by the Gillespie algorithm.

Since some processes, such as gene transcription, involve many reactions and could not be correctly modeled as an instantaneous reaction in a single step, it was proposed to model these reactions as single step multiple delayed reactions in order to account for the time it takes for the entire process to be complete.

From here, a set of reactions were proposed that allow generating GRNs. These are then simulated using a modified version of the Gillespie algorithm, that can simulate multiple time delayed reactions (chemical reactions where each of the products is provided a time delay that determines when will it be released in the system as a "finished product").

For example, basic transcription of a gene can be represented by the following single-step reaction (RNAP is the RNA polymerase, RBS is the RNA ribosome binding site, and Pro_i is the promoter region of gene i):

$$\text{RNAP} + \text{Pro}_i \xrightarrow{k_{i,bas}} \text{Pro}_i(\tau_i^1) + \text{RBS}_i(\tau_i^1) + \text{RNAP}(\tau_i^2)$$

Furthermore, there seems to be a trade-off between the noise in gene expression, the speed with which genes can switch, and the metabolic cost associated their functioning. More specifically, for any given level of metabolic cost, there is an optimal trade-off between noise and processing speed and increasing the metabolic cost leads to better speed-noise trade-offs.

A recent work proposed a simulator (SGNSim, *Stochastic Gene Networks Simulator*), that can model GRNs where transcription and translation are modeled as multiple time delayed events and its dynamics is driven by a stochastic simulation algorithm (SSA) able to deal with multiple time delayed events. The time delays can be drawn from several distributions and the reaction rates from complex functions or from physical parameters. SGNSim can generate ensembles of GRNs within a set of user-defined parameters, such as topology. It can also be used to model specific GRNs and systems of chemical reactions. Genetic perturbations such as gene deletions, gene over-expression, insertions, frame shift mutations can also be modeled as well.

The GRN is created from a graph with the desired topology, imposing in-degree and out-degree distributions. Gene promoter activities are affected by other genes expression products that act as inputs, in the form of monomers or combined into multimers and set as direct or indirect. Next, each direct input is assigned to an operator site and different transcription factors can be allowed, or not, to compete for the same operator site, while indirect inputs are given a target. Finally, a function is assigned to each gene, defining the gene's response to a combination of transcription factors (promoter state). The transfer functions (that is, how genes respond to a combination of inputs) can be assigned to each combination of promoter states as desired.

In other recent work, multiscale models of gene regulatory networks have been developed that focus on synthetic biology applications. Simulations have been used that model all biomolecular interactions in transcription, translation, regulation, and induction of gene regulatory networks, guiding the design of synthetic systems.

Prediction

Other work has focused on predicting the gene expression levels in a gene regulatory network. The approaches used to model gene regulatory networks have been constrained to be interpretable and, as a result, are generally simplified versions of the network. For example, Boolean networks have been used due to their simplicity and ability to handle noisy data but lose data information by having a binary representation of the genes. Also, artificial neural networks omit using a hidden layer so that they can be interpreted, losing the ability to model higher order correlations in the data. Using a model that is not constrained to be interpretable, a more accurate model can be produced. Being able to predict gene expressions more accurately provides a way to explore how drugs affect a system of genes as well as for finding which genes are interrelated in a process. This has been encouraged by the DREAM competition which promotes a competition for the best prediction algorithms. Some other recent work has used artificial neural networks with a hidden layer.

Gene Regulation at the Single Cell Level

Gene regulation is an intricate complex process, which involves genes, mRNAs and proteins that dictate cellular phenotypes and their response to external stimuli. Recent approaches employing genomics and proteomics and interactomic studies have helped probe the structure and signalling of these complex networks. However, more interesting aspects of the cell systems can be explored through the dynamics executed by these networks. To probe the dynamics of gene networks is complex because of the following reasons:

i. Simultaneous time resolved measurements of network components are required for dynamics studies and they generally are not accurate.

ii. There is variability in responses even among the cells that are genetically identical. Variation in cell parameters like cell size, stage of the cell cycle, metabolite concentration and intrinsic stochasticity of biochemical reactions inside the cell compel the cells to a differential response to the same external stimuli. Such variations in dynamics are often measured by distributions of relevant observations in a population which can be measured by techniques such as flow cytometry.

iii. Avoiding ambiguity in interpreting dynamic data requires tracking the fates of individual cells in the population for a time span greater than the process time. Since this time span equals several cell division cycles, collection and interpreting data over cell division cycles is a predominant challenge.

Visualization of gene expression and single cell measurements at the intracellular and intercellular levels at high resolution have made possible to understand the dynamics of gene regulation in these circuits. Imaging mRNA activity using FISH, MS2-GFP, molecular beacons, FRET and fluorescent microscopy at single molecule resolution have given us a handle to understand the gene expression at single cell level. The stochastic nature of gene expression and regulation are well studied using engineered gene networks.

Newer and evolving techniques in the field have enabled real time measurement of RNA. A major difficulty in using single fused fluorescence proteins is that they sometimes fluoresce more than the cellular auto fluorescence. The MS2 tagging system and RNA reporter RNA plasmid helps address this question. Once both the proteins are expressed in live cells the multiple fluorophores fuse to MS2 capsid proteins, bind to the MS2 binding sites in the UTR of RNA of interest. This technique gives a strong fluorescent signal, which allows visualization of an individual RNA molecule. Real time tracking of target RNA's can also be done using in vivo hybridization with molecular beacons. Molecular beacons are single stranded nucleic acid probes which contain a fluorophore and a quencher that woo apart upon binding to target RNA sequence.

Interpreting Protein Dynamics

The green fluorescence protein (GFP) and its derivative proteins facilitate real time visualization of proteins. Multi coloured fluorescence microscopy enables simultaneous measurement of multiple protein concentrations and the relative roles of intrinsic and extrinsic noise in gene expression. Fluorescence Resonance Energy Transfer (FRET) utilizes a donor and acceptor fluorophore to visualize conformational changes in individual molecules. The donor fluorophore transfers energy to the acceptor when the two come together and change the wavelength of the fluorescence signal.

Fluorescence based techniques in imaging provide the flexibility to study gene expression in single cells. Such studies show fluctuations in identical cells leading to the thought that gene regulation predominantly controls cellular properties. The stochastic variation in gene expression induces switching response in single cells creating binary 'ON', 'OFF' switches. Literature on signal processing in single cells (Issacs et al., Science. 2005), documents the relation between input and output signals in engineered E.coli networks.

The relationship between the gene expression rate and the abundance of regulatory proteins in single cells has been modelled using GRF, a gene regulation function (Rosenfeld et al., Science. 2005). This study involved an engineered two-step regulatory cascade in which the gene regulatory function in individual cells where measured dynamically and simultaneously with the input output signals. Time-lapse microscopy is employed to measure the GRF with population averaging. This interesting experiment observed dynamic fluctuations of GRF in individual cells implying that stochasticity in gene expression and minor variations in parameters limit the signal transfer property of transcription networks.

Noise propagation through a three-step transcription regulatory cascade is probed by the Van Oudenaarden group in E.coli systems. These experiments investigated the abundance of gene products as different steps in the cascade and attempted to correlate these evidences in single cell. This is done by measuring the expression changes in the input and output genes by varying the concentration of the inducers in the circuit.

Gene expression variations in individual cells may also influence the physiological states with the endogenous pathways (Acar et al., Nature. 2005). The galactose regulatory pathway in Saccharomyces cerevesiae was used as a model system to investigate the determinants of stability in cellular memory. Constructed signalling networks in E. coli, have been shown to generate bistability. Such networks have also been shown to store memory through the creation of discrete states.

Spatio Temporal Analysis of Gene Expression using Single Molecule Techniques

While fluorophore tags enable detection of target molecules and their localization in vivo fluorescence microscopy is widely adapted to determine the mean fluorescence of the tagged proteins and also to locate single molecules. The high diffusion rate of single molecules inside live cells makes their imaging difficult. This problem can be overcome by reducing their diffusion rate by localizing the molecules to the membrane where they diffuse slower than in the cytoplasm. A focused laser beam in a confocal microscope achieves single molecule detection near field scanning optical microscopy (NSOM), stochastic optical reconstruction microscopy, photo activated localization microscopy and stimulated emission and depression (STED) are the most powerful techniques for sub diffraction microscopy. These require scanning of large areas with a small window and hence are too slow for characterizing live cell dynamics. In order to explain the network dynamics completely, a mechanistic understanding of the networks is important. This approach requires high Spatio temporal resolution of fluorescent imaging and simultaneous measurement of the expression of network components because study of the average properties of cell cultures is not enough and more important is the sources of fluctuation and inter cellular variability. Micro fluidics and optogenetics help track multiple species of RNA and proteins in vivo with sub cellular evolution. Mechanical stimuli can regulate gene expression in live cells. The response of an individual cell modification to a defined molecular environment depends on the cell type and the mechanical stimulus. The mechanical stimuli require to elicit gene expression in live cells have been probed using AFM, optical tweezers and magnetic beads.

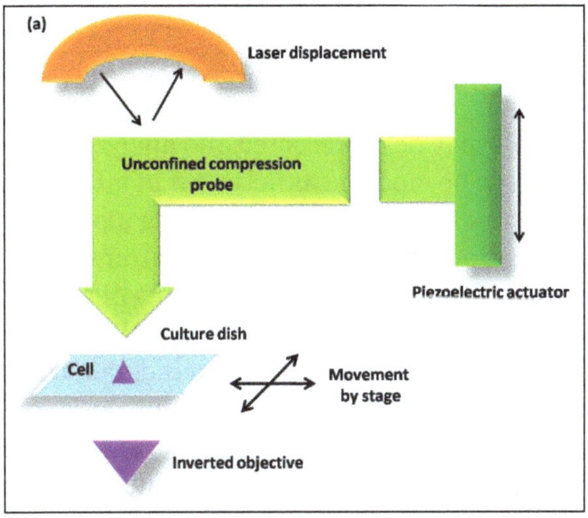

This device exposes single adherent cells to unconfined creep compression at forces ranging from 10nN to 200nN or grater. Bean theory was used to calculate the force while measuring the resulting cellular deformation.

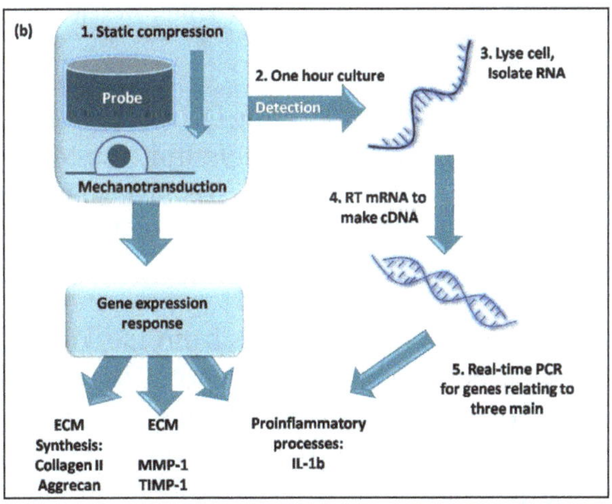

Unconfined compressions of single cells modify cell gene expression by mechanotransduction. For chondrocytes, this modification could occur in genes related to major processes involved in cartilage homeostasis and disease. Gene expression was measured in compressed chondrocytes through single-cell RT-PCR.

Biological Network

A biological network is any network that applies to biological systems. A network is any system with sub-units that are linked into a whole, such as species units linked into a whole food web. Biological networks provide a mathematical representation of connections found in ecological, evolutionary, and physiological studies, such as neural networks. The analysis of biological networks with respect to human diseases has led to the field of network medicine.

Network Biology and Bioinformatics

Complex biological systems may be represented and analyzed as computable networks. For example, ecosystems can be modeled as networks of interacting species or a protein can be modeled as a network of amino acids. Breaking a protein down farther, amino acids can be represented as a network of connected atoms, such as carbon, nitrogen, and oxygen. Nodes and edges are the basic components of a network. Nodes represent units in the network, while edges represent the interactions between the units. Nodes can represent a wide-array of biological units, from individual organisms to individual neurons in the brain. Two important properties of a network are degree and betweenness centrality. Degree (or connectivity, a distinct usage from that used in graph theory) is the number of edges that connect a node, while betweenness is a measure of how central a node is in a network. Nodes with high betweenness essentially serve as bridges between different portions of the network (i.e. interactions must pass through this node to reach other portions of the network). In social networks, nodes with high degree or high betweenness may play important roles in the overall composition of a network.

As early as the 1980s, researchers started viewing DNA or genomes as the dynamic storage of a language system with precise computable finite states represented as a finite state machine. Recent complex systems research has also suggested some far-reaching commonality in the organization of information in problems from biology, computer science, and physics, such as the Bose–Einstein condensate (a special state of matter).

Bioinformatics has increasingly shifted its focus from individual genes, proteins, and search algorithms to large-scale networks often denoted as -omes such as biome, interactome, genome and proteome. Such theoretical studies have revealed that biological networks share many features with other networks such as the Internet or social networks, e.g. their network topology.

Networks in Biology

Protein–Protein Interaction Networks

Many protein–protein interactions (PPIs) in a cell form *protein interaction networks* (PINs) where proteins are *nodes* and their interactions are *edges*. PINs are the most intensely analyzed networks in biology. There are dozens of PPI detection methods to identify such interactions. The yeast two-hybrid system is a commonly used experimental technique for the study of binary interactions.

Recent studies have indicated conservation of molecular networks through deep evolutionary time. Moreover, it has been discovered that proteins with high degrees of connectedness are more likely to be essential for survival than proteins with lesser degrees. This suggests that the overall composition of the network (not simply interactions between protein pairs) is important for the overall functioning of an organism.

Gene Regulatory Networks (DNA–Protein Interaction Networks)

The activity of genes is regulated by transcription factors, proteins that typically bind to DNA. Most transcription factors bind to multiple binding sites in a genome. As a result, all cells have complex gene regulatory networks. For instance, the human genome encodes on the order of 1,400 DNA-binding transcription factors that regulate the expression of more than 20,000 human genes. Technologies to study gene regulatory networks include ChIP-chip, ChIP-seq, CliP-seq, and others.

Gene Co-expression Networks (Transcript–Transcript Association Networks)

Gene co-expression networks can be interpreted as association networks between variables that measure transcript abundances. These networks have been used to provide a systems biologic analysis of DNA microarray data, RNA-seq data, miRNA data etc. weighted gene co-expression network analysis is widely used to identify co-expression modules and intramodular hub genes. Co-expression modules may correspond to cell types or pathways. Highly connected intramodular hubs can be interpreted as representatives of their respective module.

Metabolic Networks

The chemical compounds of a living cell are connected by biochemical reactions which convert one compound into another. The reactions are catalyzed by enzymes. Thus, all compounds in a cell are parts of an intricate biochemical network of reactions which is called metabolic network. It is possible to use network analyses to infer how selection acts on metabolic pathways.

Signaling Networks

Signals are transduced within cells or in between cells and thus form complex signaling networks. For instance, in the MAPK/ERK pathway is transduced from the cell surface to the cell nucleus by a series of protein–protein interactions, phosphorylation reactions, and other events. Signaling networks typically integrate protein–protein interaction networks, gene regulatory networks, and metabolic networks.

Neuronal Networks

The complex interactions in the brain make it a perfect candidate to apply network theory. Neurons in the brain are deeply connected with one another and this results in complex networks being present in the structural and functional aspects of the brain. For instance, small-world network properties have been demonstrated in connections between cortical areas of the primate brain. This suggests that cortical areas of the brain are not directly interacting with each other, but most areas can be reached from all others through only a few interactions.

Food Webs

All organisms are connected to each other through feeding interactions. That is, if a species eats or is eaten by another species, they are connected in an intricate food web of predator and prey interactions. The stability of these interactions has been a long-standing question in ecology. That is to say, if certain individuals are removed, what happens to the network (i.e. does it collapse or adapt)? Network analysis can be used to explore food web stability and determine if certain network properties result in more stable networks. Moreover, network analysis can be used to determine how selective removals of species will influence the food web as a whole. This is especially important considering the potential species loss due to global climate change.

Between-species Interaction Networks

In biology, pairwise interactions have historically been the focus of intense study. With the recent advances in network science, it has become possible to scale up pairwise interactions to include individuals of many species involved in many sets of interactions to understand the structure and function of larger ecological networks. The use of network analysis can allow for both the discovery and understanding how these complex interactions link together within the system's network, a property which has previously been overlooked. This powerful tool allows for the study of various types of interactions (from competitive to cooperative) using the same general framework. For example, plant-pollinator interactions are mutually beneficial and often involve many different species of pollinators as well as many different species of plants. These interactions are critical to plant reproduction and thus the accumulation of resources at the base of the food chain for primary consumers, yet these interaction networks are threatened by anthropogenic change. The use of network analysis can illuminate how pollination networks work and may in turn inform conservation efforts. Within pollination networks, nestedness (i.e., specialists interact with a subset of species that generalists interact with), redundancy (i.e., most plants are pollinated by many pollinators), and modularity play a large role in network stability. These network properties may actually work to slow the spread of disturbance effects through the system and potentially buffer the pollination network from anthropogenic changes somewhat. More generally, the structure of species interactions within an ecological network can tell us something about the diversity, richness, and robustness of the network. Researchers can even compare current constructions of species interactions networks with historical reconstructions of ancient networks to determine how networks have changed over time. Recent research into these complex species interactions networks is highly concerned with understanding what factors (e.g., diversity) lead to network stability.

Within-species Interaction Networks

Network analysis provides the ability to quantify associations between individuals, which makes it possible to infer details about the network as a whole at the species and/or population level. Researchers interested in animal behavior across a multitude of taxa, from insects to primates, are starting to incorporate network analysis into their research. Researchers interested in social insects (e.g., ants and bees) have used network analyses to better understand division of labor, task allocation, and foraging optimization within colonies; Other researchers are interested in how certain network properties at the group and/or population level can explain individual level behaviors. For instance, a study on wire-tailed manakins (a small passerine bird) found that a male's degree in the network largely predicted the ability of the male to rise in the social hierarchy (i.e. eventually obtain a territory and matings). In bottlenose dolphin groups, an individual's degree and betweenness centrality values may predict whether or not that individual will exhibit certain behaviors, like the use of side flopping and upside-down lobtailing to lead group traveling efforts; individuals with high betweenness values are more connected and can obtain more information, and thus are better suited to lead group travel and therefore tend to exhibit these signaling behaviors more than other group members. Network analysis can also be used to describe the social organization within a species more generally, which frequently reveals important proximate mechanisms promoting the use of certain behavioral strategies. These descriptions are frequently linked to ecological properties (e.g., resource distribution). For example, network analyses revealed subtle differences in the group dynamics of two related equid fission-fusion species, Grevy's zebra and onagers, living in variable environments; Grevy's zebras show distinct preferences in their association choices when they fission into smaller groups, whereas onagers do not. Similarly, researchers interested in primates have also utilized network analyses to compare social organizations across the diverse primate order, suggesting that using network measures (such as centrality, assortativity, modularity, and betweenness) may be useful in terms of explaining the types of social behaviors we see within certain groups and not others. Finally, social network analysis can also reveal important fluctuations in animal behaviors across changing environments. For example, network analyses in female chacma baboons (*Papio hamadryas ursinus*) revealed important dynamic changes across seasons which were previously unknown; instead of creating stable, long-lasting social bonds with friends, baboons were found to exhibit more variable relationships which were dependent on short-term contingencies related to group level dynamics as well as environmental variability. This is a very small set of broad examples of how researchers can use network analysis to study animal behavior. Research in this area is currently expanding very rapidly. Social network analysis is a valuable tool for studying animal behavior across all animal species, and has the potential to uncover new information about animal behavior and social ecology that was previously poorly understood.

Network Motifs

Let us try to decipher the complex networking inside cells now. Every living cell is comprised of intricate networks that represent interactions and inter dependent events among the bio molecules in the network. Gene regulation networks, protein- protein interaction networks, metabolic pathways, and neural networks are well known examples of this intricacy.

The primary aim of bio medical research in the post genomic era has been to catalogue all the bio molecules and their interactions within a living cell. Advances in the field of network biology such as high throughput data collection techniques, micro arrays, protein chips, automated yeast two hybrid strains permit simultaneous investigation of cellular components and information on the molecular interaction between these components.

These networks are not independent but instead form a 'network of networks' which regulate cellular behavior. The intricate complexity of the network poses a major challenge to contemporary biology and demands an integrated theoretical and experimental approach that can map out and model the topological and dynamic properties of these networks.

This class will focus on tools which can quantify network parameters and therefore help us understand the cell's internal organization and evolution. I am sure this will completely alter our perspective of a cell and the cellular mechanisms in place. Understanding the organising principles of cellular network leads to a better understanding of the cell as a system and also provides relevance for experimental biologists in the role of individual molecules in various cellular processes.

Cellular Networks-Nomenclatures

All of us know that complex systems ranging from the internet to a living cell behave due to the orchestrated activity of network components that show pair wise and more complex interactions. These components can be reduced to a series of 'nodes' connected to each other by certain 'links', each link representing the interaction between two components. These nodes and links constitute a network. This is also called 'a graph' in the mathematical language. Physical interactions such as protein- protein interactions, protein-nucleic acid interactions and protein metabolites interactions can be concretely explained with the node link nomenclature.

Networks can be classified as 'directed' or 'undirected' based on the nature of the interaction. Figure illustrates this through an elegant graphical representation.

Directed Network

In such networks as shown in Fig (a) the interaction between any two nodes has a well defined direction like the direction of signalling from a transcription factor to a gene or the direction of material flow from a substrate to a product in a metabolic reaction.

Undirected Network

These are networks as shown in Fig (b) in which the links have not been assigned a specific direction.eg. In protein interact networks a link represents a mutual binding relationship. If protein A binds to protein B then protein B also binds to protein A.

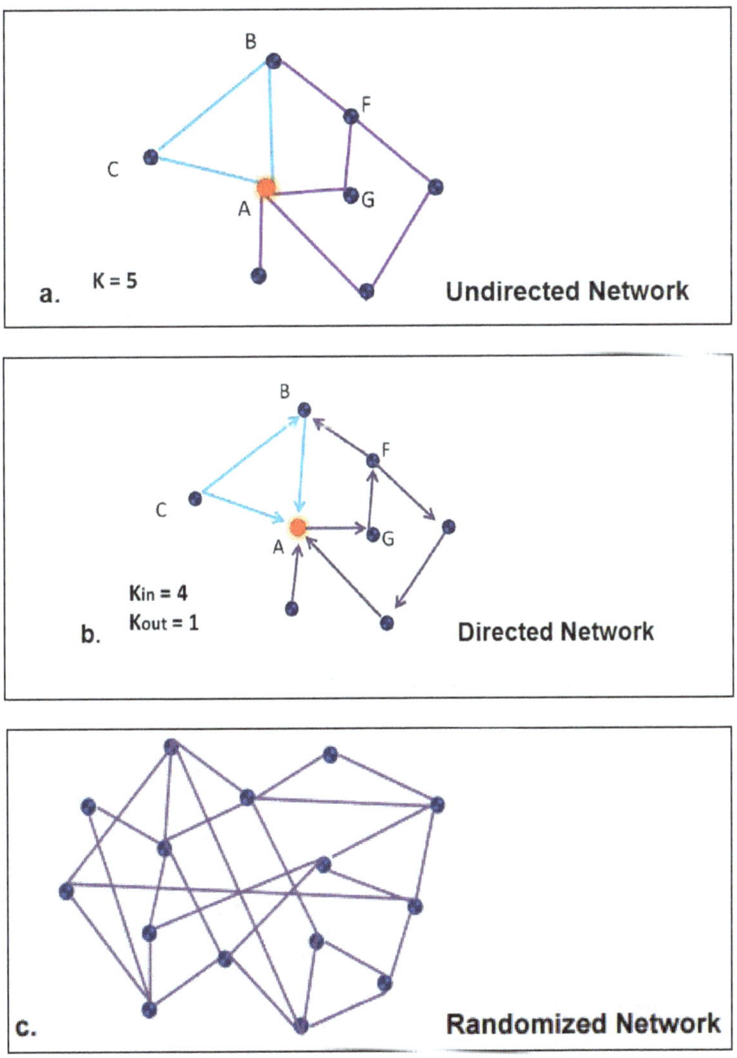

Representation of a biological network.

Patterns, Randomized Networks and Network Motifs in Biology

Network motifs are simple building blocks of complex networks and are statistically over represented sub structures or sub graphs in a network. Since the number of sub graphs in biological networks increases exponentially with the network or motif size, it is difficult to detect larger network motifs in a biological network. Hence network motifs can be defined as recurring patterns of interactions that are significantly over represented in a biological system. This over representation of the sub network indicates the functional importance of such motif. Therefore it becomes important to explore these abundant motifs in biological networks. Milo et al, (Science 2002) in one of their break through explorations, work analyzed 18 different networks from,

1. Transcription networks of E.coli and S.cerevisiae.
2. Synaptic connections between neurons in c. elegans.
3. Throphic interactions between predator and prey in ecological systems.

As represented in Fig (a) in each of these networks the number of nodes is represented by 'n'. In each of the above mentioned cases all possible motifs of size n=3 and n=4 were enumerated and compared to an average count over thousand random networks. Randomized networks were generated without compromising on certain properties of the original network.

1. In-degree, out- degree and mutual degree: this is done by swapping edges to generate random graphs.
2. The number of appearances of all n-1 node sub graphs for n>3. This is done to ensure that high significance is assigned to highly significant sub pattern also.

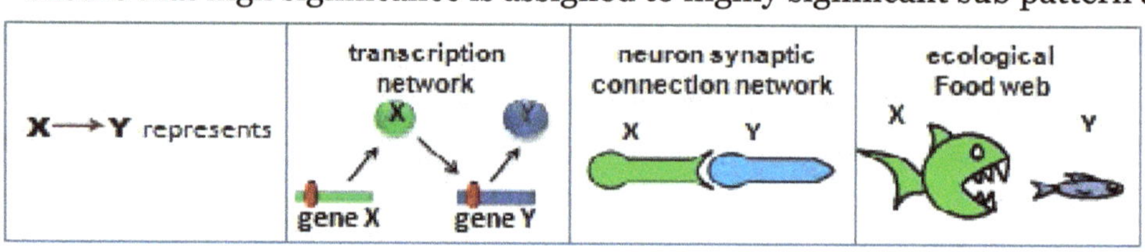

(a) Types of network investigated in Milo et. al (Science 2002).

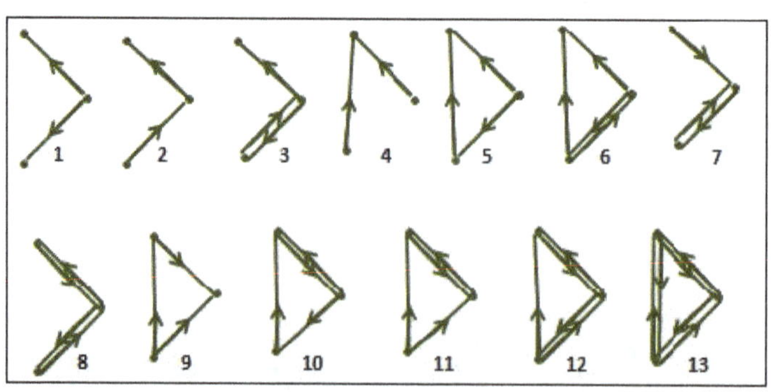

(b) Directed motifs for n=3 where n refers to the number of nodes.

Identification of Network Motifs

The following rules help us identify network motifs in biological systems:

1. Identification all sub graphs of n nodes in the network.
2. Randomization of the network without changing the number of nodes, edges and degree distribution.
3. Identification of all sub graphs of n nodes in the randomized network.
4. Comparison of more frequently occurring significant sub graphs with randomized ones in the network and designating them as motifs.

Real live transcription networks of organisms like E.coli show numerous patterns of nodes and edges but it is important to look for meaningful patterns that are statistically significant to derive biological information that brings us to discuss the concept of randomized networks.

Randomized networks are a type of networks that posses the same characteristics of a real network. These have the same numbers of nodes and edges as in biological systems. But in these networks random connections are made between nodes and edges. Network motifs are patterns that occur more significantly and more often in real networks than in randomized networks. Recurrent patterns give us the basic idea that these must have been evolutionarily conserved against mutation.

The best way to illustrate this is the fact that edges are easily lost in a transcription network and a single mutation abolishes transcription factor binding hence facilitating the loss of edge in the network. In the same way mutations which generate a binding site for transcription factor X in promoter region of gene Y can help add new edges to the network. Mutations, duplication events that reposition pieces on the genome or insertion events can generate new binding sites and hence add new edges to the network. This clearly demonstrates that the occurrence of the network motifs more often than in randomized network proves that this selection offers an advantage to the organism.

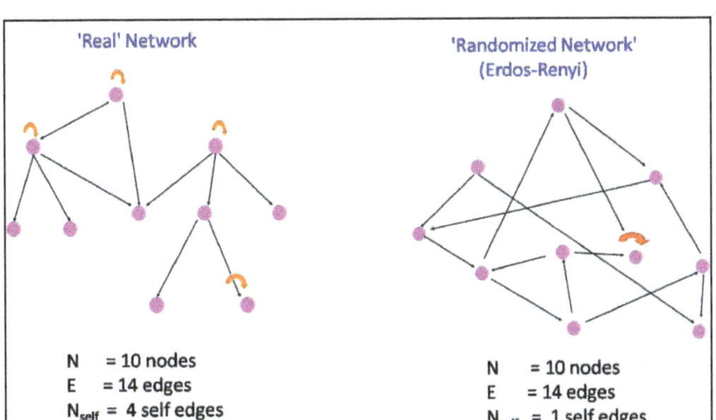

Representation of Real and Randomized Network with nodes, edges and self edges.

Auto Regulation

Self Edges

Self edges are those edges that originate and end at the same node. The *E.coli* network has approximately 40 self edges each of which correspond to transcription factors that regulate the transcription of their own genes. This type of regulation of a gene by its own gene product is called autogenous control or auto regulation. 34 of the auto regulatory proteins in the *E.coli* network have been found to repress their own transcription. This process is referred to as negative auto regulation as referred to in Figure.

Negative auto regulation is a network motif and occurs at higher numbers than expected in random networks. Such structures display engineering advantages and are more prone to evolutionary selection in their appearances network motif.

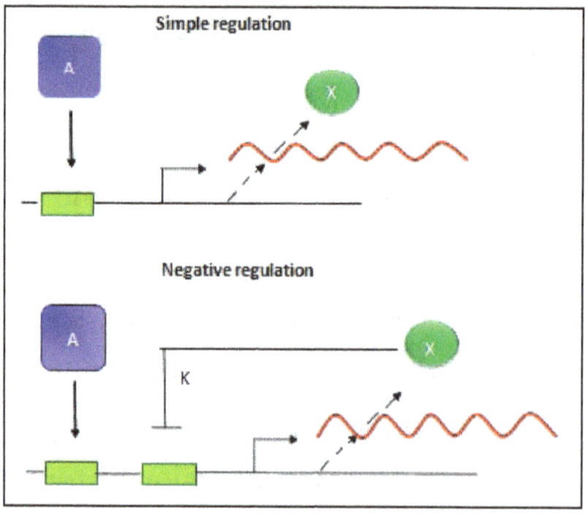

Schematic representation of Simple and Negative Auto Regulation in Biological Networks.

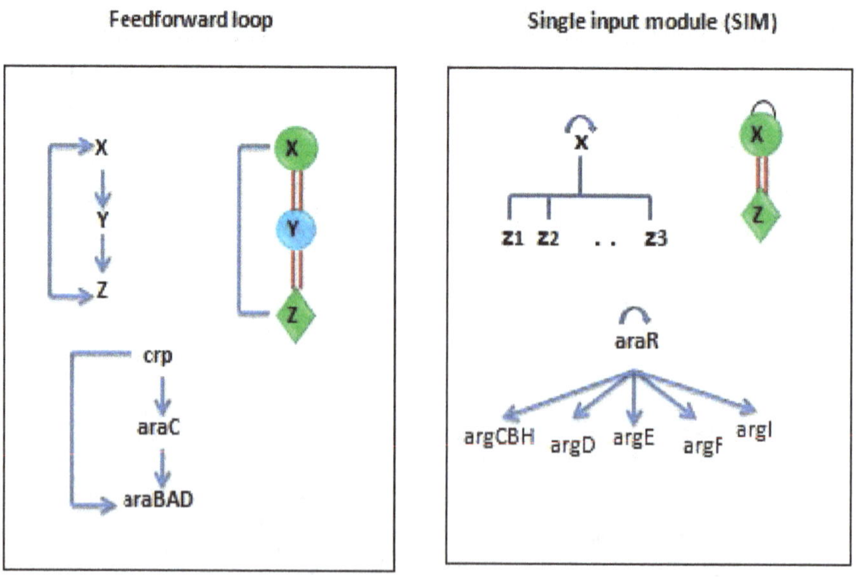

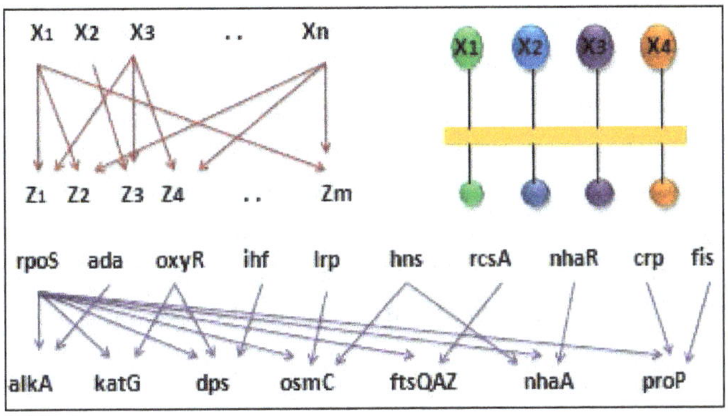

Dense overlapping regulons.

Feed Forward Loops

In order to understand the general features of such networks and to extract useful information from them, we dissect them at hierarchical levels- into modules and motifs which can explain their functionality, evolution and dynamic behavior. Over the process of evolution, these networks show information processing functions. Interesting investigations on network behaviour have shown that simple switching circuits, amplifiers or oscillators can map to the core process of biological decision making. These have been implemented by two or three gene network motifs and are characterized by how they behave around fixed points in the system. Here the steady state of the system as well as the process of achieving equilibrium in the system reflects the characteristic function performed by the genetic circuit.

Network motifs appear at frequencies much higher than those expected at random and hence imply information processing roles for these motifs. To arrive at such significant patterns, one first identifies the different patterns of these motifs in real and randomized networks and then calculates the number of appearances of these patterns in the real and random networks. The discussion that follows focuses on patterns with 3 nodes (forming a triangle), Figure below. There are 13 possible 3-node patterns in such arrangement. Of these, only one of them qualifies to be a network motif called the Feed Forward Loop.

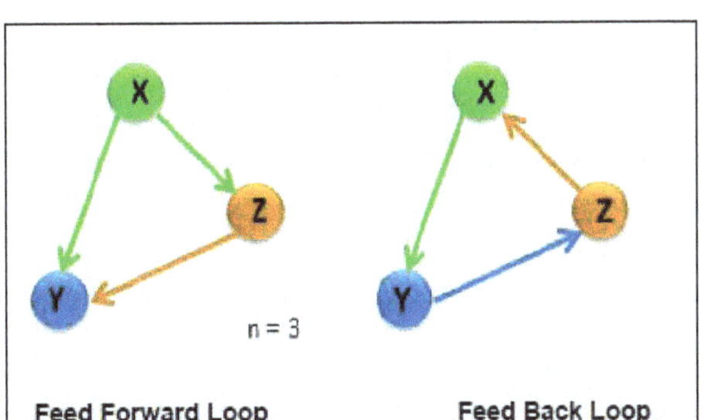

Representative Feed Forward and Feed Back Loops with nodes n=3 forming a triangle.

The most significant of the network motifs in E. coli and yeast is the Feed Forward Loop which is defined by a transcription factor X that regulates a second transcription factor Y. X and Y both jointly regulate an operon Z by binding to its regulatory region. Here X is called the general transcription factor, Y the specific transcription factor and Z the effector operon. As described in Figure, this type of motif occurs in the L-arabinose utilization system where Crp is the general transcription factor and Ara-C is a specific transcription factor. Such a motif characterizes 40 effector operons in 22 different systems in the network database and accommodates 10 different transcription factors.

Feed Forward Loop

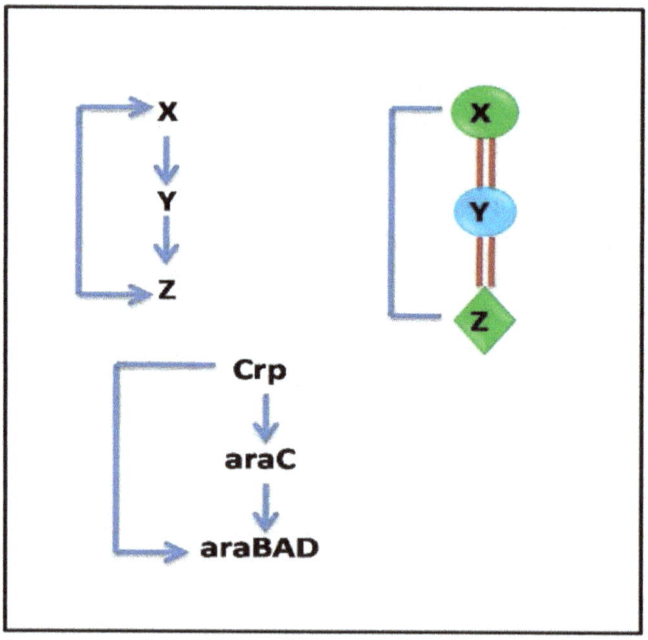

Diagrammatic representation of FFL with an example of L-arabinose utilization system.

In order to appreciate the dynamics of the Feed Forward Loop, one needs to study the regulation at each one of its three edges, each of which can represent activation or a repressive interaction. S_X and S_Y are the two input signals to the Feed Forward Loop. The signals could be small molecules, protein partners, biochemical or environmental stimuli or covalent modifications that activate or inhibit the transcription of X and Y. Therefore there are 8 possible types of FFL - 8 structural configurations of activator and repressor interactions.

Coherent and Incoherent Feed Forward Loops

A Feed Forward Loop is termed 'coherent' if the direct effect of the general transcription factor X on the effector operons has the same sign (positive or negative) as its net indirect effect through the specific transcription factor.

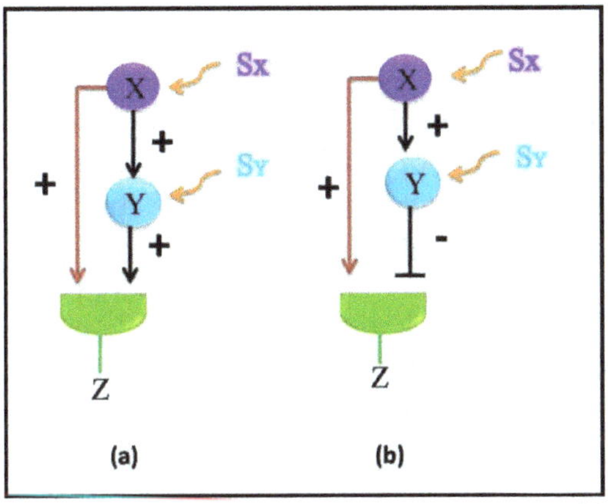

Feed-Forward Loop (FFL) (a) Coherent FFL (type1), X regulates Y and both jointly regulate Z. (b) Incoherent FFL (type1), here Sx and Sy are the signals.

In other words, if X regulates Y positively and if X and Y both positively regulate Z, the Feed Forward Loop is coherent. i.e. the sign of the direct path of regulation (X to Z) is the same as the overall sign of the indirect regulation path (X to Z through Y). The overall sign of a path is determined by the multiplication of the sign of each arrow on the path. In Figure (a) we see that the sign of the indirect path (X→Y→Z) is plus x plus =plus, while the direct path (X→Z) is already plus. Since both the direct and indirect paths have the same positive sign, this loop is called a Coherent Feed Forward Loop.

The other type of FFL is called Incoherent FFL in which the sign of the indirect path of regulation is opposite to that of the direct path. In type-1 Incoherent FFL as denoted the direct path is positive and the indirect path is negative. The Incoherent FFLs show odd number of minus edges.

In both the coherent and incoherent loops, the effects of the general and specific transcription factors X and Y are integrated at the promoter region of gene Z. The expression profile of Z is modulated by the concentrations of X and Y bound to their inducers. The cis regulatory input function of Z describes this modulation. cis regulatory input functions include logic gates like AND which require both X and Y to express Z and OR gates in which either X or Y is sufficient to express Z.

Both Coherent and Incoherent Feed Forward Loops are sign sensitive. Type 1 coherent FFLs (in which all three regulations are positive) are the most abundant type of Feed

The Incoherent Feed Forward loop type-1 is the second most abundant type of FFL among biological networks. The other types of feedforward loop do not appear more frequently than CFFL I and ICFFL I.

Coherent Feed Forward Loops

The Feed Forward Loop is a network motif constituted by a transcription factor X which regulates another transcription factor Y and both X and Y regulate a gene

Z. Such Feed Forward Loops comprise two parallel paths that regulate the circuit, a direct path from X->Z and an indirect path of X->Z through Y. A Coherent Feed Forward Loop has the same overall sign for the indirect path as that of the direct path of regulation. Though all the Feed Forward Loops appear with equal frequency across transcription networks, Type -1 Coherent FFL is the most abundant type of FFL and shows a positive sign for all the regulations involved in the circuit.

In order to explain the functional logic exhibited by the Feed Forward Loop, let us consider a typical situation inside a cell during the process of transcription. Let the cell express multiple copies of a protein X, one of the transcription factors in the Feed Forward Loop. Let signal S_x be the input to X. This X is inactive without the signal S_x. At a time t=0, the strong signal S_x triggers the activation of X, transiting it to a form X^*. This process is called a step like stimulation of X.

X^*, the active protein now binds to the promoter of gene Y initiates production of protein Y, the second transcription factor in the Feed Forward Loop. Mean while X^* binds to the promoter of gene Z. If the circuit follows an AND logic (input function at the Z promoter), X^* alone cannot activate the production of Z. it requires the binding of both X^* and Y^*. This indicates that Y should build up to sufficient levels of expression to cross the activation threshold of gene Z (K_{YZ}).Activation of Z also requires the presence of S_y which activates Y to its form Y^*. Therefore the appearance of the signal S_x and Y are both required to activate Z. This introduces a delay in the production of Z. This function is represented by the truth tables below.

Coherent Feed Forward Loop Type I (CFFL I)

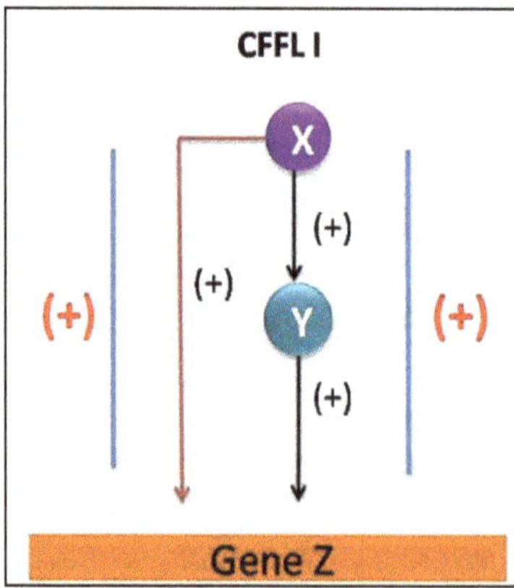

Z Logic	AND	OR
Steady- state		
Z(Sx, SY)	Sx ^ SY	Sx
Response delay		
Sx ON step	Delay	--
Sx OFF step	--	Delay
Inverted out	No	No

(a) Coherent Feed Forward Loop.

Table (a) AND- and OR- gates at the Z Type I (CFFL I) promoter for CFFL I.

Case 1: Steady State

In an AND logic circuit, described in Figure (a) when both the signals S_X and S_Y are present. In the presence of signals the system reaches a steady state. The active part of transcription factor X which is X* binds to the promoter gene Z through gene Y and X binds to Z independent of Y. So gene Z receives two inputs (one from gene X and the other from Y) leading to the expression of both the genes.

In short, X controls Z, forming a direct pathway. X controls Z, through Y forming an indirect pathway.

Case 2: S_X ON State

Let us now consider the case when S_X is ON. In the AND circuit, When SX is on, the transcription factor of the protein X becomes active X*and binds to the promoter of gene Z, thus expressing Z. (direct path).

The activated protein X* also binds to promoter of the gene Y and activates Y (Y*) as a result of which gene Y gets expressed. Since Y should reach the activation threshold, to bind to the promoter of gene Z (indirect path), the process requires time. Hence delaying the expression of Z.

Case 3: S_x OFF State

In the AND circuit, when S_x is OFF, it cannot trigger the expression of Z directlu or through Y. Hence there is no expression of gene Z.

In an OR function, when S_x is OFF, the transcription factor of the protein X does not become active X* and does not bind to the promoter of gene Z and gene Z is not expressed(direct path). The protein X does not binds to promoter of the gene Y and gene Y is not fully expressed. When Y reaches the activation threshold, it binds to the promoter of gene Z (indirect path). Since the levels of expression of Y isles in this case, the process is delayed and so is the expression of gene Z.

Case 4: Inverted Out

In both AND circuit, when the system is inverted, this case results in no expression of Z. Similarly OR gate can be worked out for all the four cases.

Coherent Feed Forward Loop type II (CFFL II)

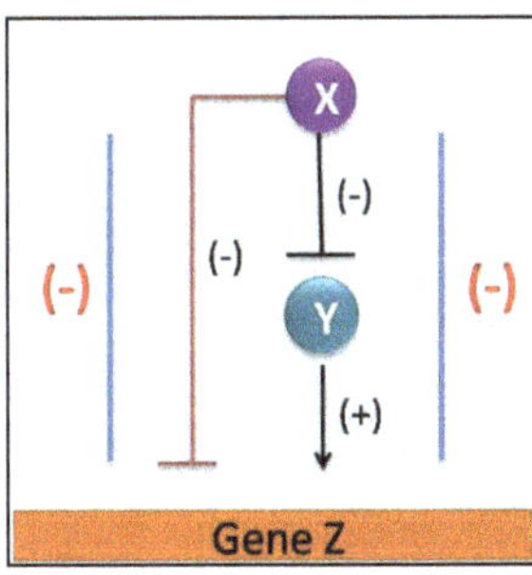

(b) Coherent Feed Forward Loop.

CFFL II

Z Logic	AND	OR
Steady- state		
Z(S_X, S_Y)	$\hat{S}_X \wedge S_Y$	$\hat{S}_X$
Response delay		
Sx ON step	--	Delay
SX OFF step	Delay	--
Inverted out	Yes	Yes

Table (b) AND- and OR- gates at the Z Type II (CFFL II) promoter for CFFL II.

Case 1: Steady state

In the AND circuit, the system reaches a steady state when Sx$^\wedge$ (the inverted input signal of S_X) and signal S_Y are present. In the presence of Sx$^\wedge$ signal, the active part of the transcription factor X which is X^* binds to the promoter of gene Z through gene Y and also binds to gene Z independent of gene Y. Hence both the genes are expressed in this case.

In an OR circuit, the system requires only the inverted signal Sx (Sx$^\wedge$) to reach steady state.

Case 2: S_X ON State

In an AND circuit, when S_X is ON, there is no gene expression.

The activated protein X* also binds to promoter of the gene Y and represses gene Y. But Y reaches the activation threshold K and binds to the promoter of gene Z (indirect path). Since this process requires time for Y to reach the threshold. The expression of gene Z is delayed.

Case 3: S_X OFF State

In the AND circuit, when S_X is OFF, the transcription factor of the protein X does not get activated X* and hence does not bind to the promoter of gene Z. Hence is not repressed (direct path).

The transcription factor of the protein X does not bind to promoter of the gene Y and gene Y is also not repressed. Y reaches the activation threshold, it binds to the promoter of gene Z (indirect path). Since this process depends on the activation of Y and the time required to reach the threshold. The expression of gene Z is delayed.

Case 4: Inverted Out

In both AND circuit, when the system is inverted, there is expression of gene Z without any delay, as its gene expression is controlled by the active X*(direct path).

Similarly OR circuit can be worked out for all the four cases.

Coherent Feed Forward Loop type III (CFFL III)

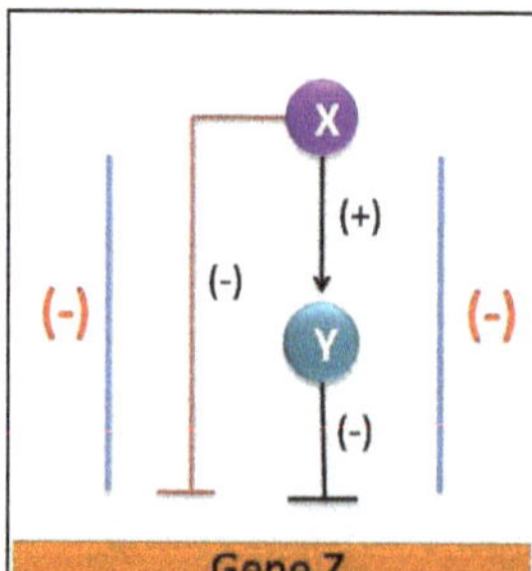

(c) Coherent Feed Forward Loop.

CFFL III

Z Logic	AND	OR
Steady- state		
Z(Sx, Sy)	Ŝx	Ŝx ^ Ŝy
Response delay		
Sx ON step	--	--
Sx OFF step	Delay	Delay
Inverted out	Yes	Yes

Table (c) AND- and OR- gates at the Z Type III (CFFL III) promoter for CFFL III.

Case 1: Steady State:

In the AND circuit, the system reaches steady state when the signals inverted S_X is present. In the presence of inverted Sx signal, the active part of the transcription factor of X which is X^* binds to the promoter of gene Z.

Case 2: S_X ON State

In an AND, when Sx is ON, there is no gene expression of Z.

Case 3: S_X OFF State

In an AND circuit, when S_X is OFF, the transcription factor of the protein X cannot make an active X*and hence does not bind to the promoter of gene Z and gene Z is not repressed (direct path).

The transcription factor of the protein X does not bind to promoter of the gene Y and gene Y is also not expressed. Hence gene Y will not repress gene Z because there is no production of Y (indirect path). The expression of gene Z is delayed.

Case 4: Inverted Out

In both AND circuit and OR circuit, when the system is inverted, there is expression of gene Z with delay, as its gene expression is controlled only by the active X* (direct path).

Similarly OR circuit can be worked out for all the four cases.

Coherent Feed Forward Loop Type IV (CFFL IV)

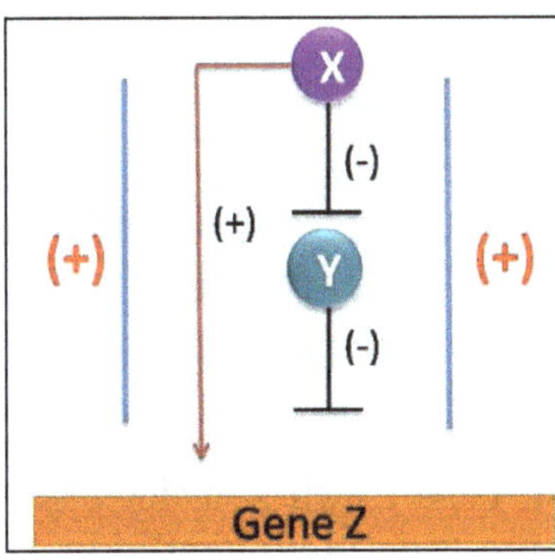

(d) Coherent Feed Forward Loop.

CFFL IV

Z Logic	AND	OR
Steady- state		
Z(Sx, SY)	Sx	Sx ˇ ŜY
Response delay		
Sx ON step	Delay	Delay
Sx OFF step	--	--
Inverted out	No	No

Table (d) AND- and OR- gates at the Z Type IV (CFFL IV) promoter for CFFL IV.

Case 1: Steady State

In an AND circuit, the system reaches steady state when the signal S_X is present. In the presence of the signal S_X, the active part of the transcription factor of X which is X^* binds to the promoter of both gene Z through gene Y and also to gene Z independent of gene Y. Therefore gene Z is expressed even when gene Y is repressed. The repression of gene Z does not take place through gene Y.

Case 2: S_X ON State

In both AND circuits, when S_X is on, the transcription factor of the protein X becomes active X*and binds to the promoter of gene Z and expresses gene Z (direct path).The activated protein X* also binds to promoter of the gene Y and represses gene Y. The repression of gene Z does not take place through Y. So the expression of gene Z is delayed.

Case 3: S_X OFF State

In both AND circuit, when S_X is OFF, there is no gene expression.

Case 4: Inverted Out

In both AND circuit, when the system is inverted, there is no expression of gene Z. Similarly OR circuit can be worked out all the four cases.

Besides the logic functions and truth tables discussed elaborately in this lesson, Coherent Feed Forward Loops have well interesting properties in biological systems. An outstanding example is the function of Type-1 AND gate Coherent FFL as a sign sensitive delay in the arabinose system in E.coli in the presence of cyclic AMP signals. We shall deal with these in the next few classes which bring out a clear understanding of the information processing roles of the biological networks.

Kinetic Response of Feed Forward Loops

The kinetic response of the Feed Forward Loops are interesting to observe in living cells because they regulate functional responses in the circuit through decision logic. We shall now try to understand how a Coherent Feed Forward Loop can be used to induce a sign dependent delay in the circuit and how such sign sensitive delays are relevant to the arabinose system of *E.coli.* We shall also discuss how an Incoherent FFL accelerates the response time in the circuit, tracing the example of the galactose system of *E.coli.*

In the earlier class, we learnt how the Coherent FFL Type-1 with AND logic introduces a delay when the signal S_X is ON. Let us recall that in the Type-1 Coherent FFL (AND gate), when S_X takes a step addition, the activator Y grows to a sufficient concentration and crosses the activation threshold. The expression of Z begins only now. Therefore there is a delay introduced in the circuit. This behavior is called sign sensitive delay in the circuit because the delay is dependent on whether the step is ON or OFF. The speed of the response of the circuit is characterized by the term 'response time', defined as the time taken by the gene product Z to reach 50% of its physiologically determined steady state level of expression. Let us consider a circuit where the pulse S_X (ON pulse) appears for brief time.

When S_X is suddenly removed following an OFF step, S_X becomes inactive, unbinds from the promoters Y and Z. therefore the AND gate requires only one of its inputs to go OFF in order to stop the expression of Z. Thus an OFF step causes no delay in the dynamics of Z.

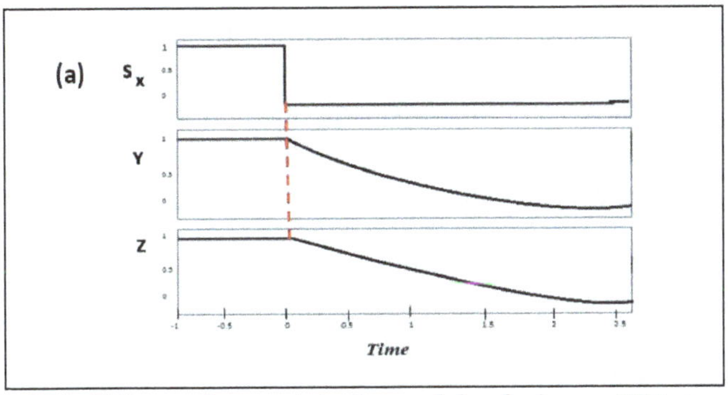

The C1-FFL AND gate doesnot introduce a delay during an OFF step of S_X
The Fig depicts the dynamics of the FFL during an OFF step of SX at time t=0.

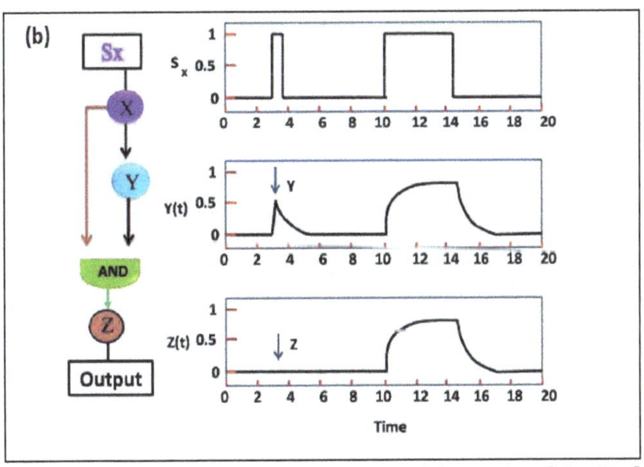

The C1-FFL AND gate behaves as persistence detector. On receiving a signal S_X, Y takes time to accumulate and cross the activation threshold to activate Z. Since the persistent pulse causes a delay in the production of Z, the circuit is called a persistent detector circuit.

Persistent Detector

If the input ON pulse is shorter than the delay time TON, the input will not result in the expression of Z in the C-1 FFL circuit. This is because the gene Y cannot cross its activation threshold during the brief period of the ON pulse. This indicates that only pulse is longer than T_{ON} lead to the expression of Z. Since such persistent pulses lead to the expression of Z, this Feed Forward Loop C-1 FFL is called a 'persistent detector' as shown in Figure for ON pulses. The same circuit responds immediately to OFF pulses. The major difference in the behavior of C-1 FFL to a simple regulatory circuit which is not an FFL is that a simple regulatory circuit does not filter out short input pulses and shows expression of Z as long as the input pulse is present.

OR gate C-1 FFL Functions as a Sign Sensitive Delay Circuit During OFF Steps of S_X

Let us now replace the AND gate in the C-1 FFL with an OR gate at the Z promoter. How does the response of the circuit change? When the C-1 FFL has an OR gate, the gene Z gets activated immediately when S_X is ON. This is because one ON input is enough to activate the function of an OR gate. Hence no delay is introduced following the ON step of S_X. But Z is inactive due to a delay following an OFF step because both the inputs go OFF for the OR gate to be inactivated. Thus Y^* can activate Z even without X^*. It takes time for Y^* to decay after an OFF step of S. Therefore the Coherent.

FFL Type-1 with an OR function also behaves as a sign sensitive delay element with signs opposite to that of its AND version. Experimental measurements on gene expression at high resolution have observed such a dynamic circuit which induce a delay with S_X OFF during OR function in the flagella system of *E.coli*. The OR gate C-1 FFL has been shown to regulate the production of proteins that self assemble into a motor which facilitates *E.coli* swimming through rotational movement of the flagella. Accurate measurements under experimental conditions have shown that the delay after removal of the input S_X is around one hour (one cell generation time). The delay has the same order of magnitude as the time it takes to assemble a flagella motor. The interesting feature with the OR gate C-1 FFL is that it continues to express the gene for about an hour even after the input signal fades. This means that the OR C-1 FFL is capable of protecting the genetic circuit against transient loss of input signal.

Significance of Sign Sensitive Delay in Biological Networks

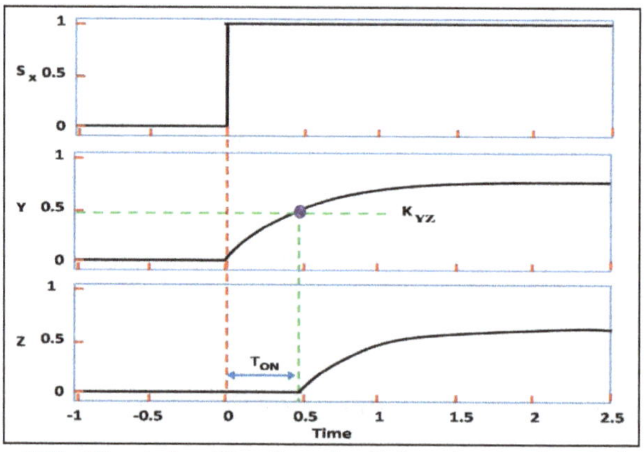

The c1-FFL OR gate is a Sign Senstive Delay circuit when S_X is OFF.

As discussed, sign sensitive delays protect circuits against brief fluctuations in input. Evolutionary demands might have positioned Type-1 CFFLs in diverse systems inside a cell which requires protection against transient loses. We know that the cellular environment is continuous and highly fluctuating system where the stimulating signals could prevail for very brief periods which may not be sufficient to elicit a response. Since the sign sensitive delay element in C-1 FFL also acts as an asymmetric filter it confers an advantageous filtering function to the fluctuating environment inside a cell. This leads us to believe that the natural selection of a Feed Forward Loop is primarily based on the filtering function we can provide to the cells.

Sign Sensitive Delay in Arabinose System

The proteins in the arabinose system transport the sugar arabinose to the cell and break it down to form an energy or carbon source. Generally, the cell prefers glucose to arabinose. Hence, the sugars arabinose and glucose are the inputs for an arabinose system to make the decision for uptake. Obviously, the prerequisite for the proteins to be formed inside the arabinose system is the presence of arabinose, and not glucose. The absence of glucose in a cell comes to light by the formation of a small molecule named cAMP. The two transcription activators, namely CRP (detects cAMP) and araC (detects arabinose) are connected in a C1-FFL with in an AND input function, as shown in Figure. Experiments in the environment of *E. coli* revealed that the delay in the arabinose FFL is on the same order of magnitude as the duration of spurious pulses of the input signal that are generally observed when *E.coli* transits between different growth conditions.

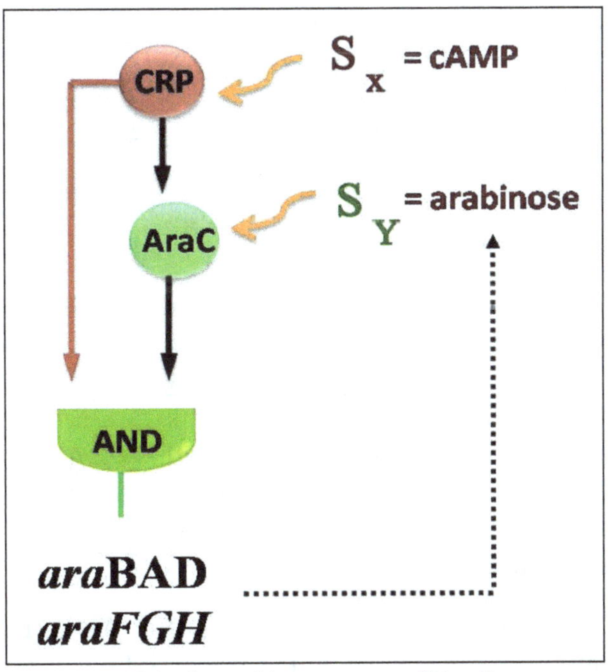

FFL in the arabinose system of E. coli.

Thus, the FFL in the system, apart from being able to filter out the short fluctuations in the input signal, responds only to persistent stimuli for instance, glucose starvation, which need to make use of sugar arabinose. Though embedded in additional interactions, the FFL in arabinose system shows sign-sensitive delay. In spite of the three-gene FFL circuit in isolation, the arabinose FFL shows the expected dynamics also when embedded within the interaction networks of the cell.

Transcription Networks

The living cell is a complex machinery involving thousands of interacting proteins and other biological molecules. These cells utilise proteins based on the situations they encounter. During sugar sensing, the cells begin to secrete proteins that transport sugar into the cells. Cells respond to damage by producing damage repair proteins. Every cell thus senses its environment continuously, regulates protein production and maintains cellular homeostasis. Live cells as we know are kept dynamic through gene expression programs that involve the transcription of thousands of genes in a coordinated manner. As we all know, the expression of a gene is facilitated by transcription regulatory proteins which recognize specific promoter sequences. The association of regulatory proteins with genes across a genome and the continuous cascade of information processing which senses the rate of production of a particular protein inside a cell constitute a transcription regulatory network. These are in general called transcription networks. Metabolic networks describe the possible pathways that a cell may use to accomplish metabolic processes. In a similar way the map of the transcriptional regulatory network of an organism describes potential pathways the cells of the organisms utilise to regulate global gene expression programs. This network map establishes a high connectivity between gene expression programs and cellular functions through networks of transcriptional regulatory molecules which in turn regulate other molecular players in transcription.

Transcription Network-Elements

We know that transcription networks depict the interaction between transcription factors and genes. The process of transcription ensures that RNA polymerase produces mRNA (messenger RNA) corresponding to the coding sequence of a gene. The mRNA is translated to form a protein which is also referred to as the gene product. Each gene is preceded by a small stretch of regulatory DNA called promoter. The promoter is a specific DNA sequence that can bind RNA polymerase, which is a complex of several proteins. The rate at which the gene is transcribed is determined by the quality of the binding site and the promoter.

which controls the number of mRNA produced per unit time. The RNA polymerase complex acts on a number of genes while the transcription factors regulate changes in expression profiles of specific genes. The transcription factors when bound change the probability per unit time of RNA polymerase binding to the promoter to produce the mRNA. Transcription factors can also function as activators which increase the transcription rate of a gene or as repressors which reduce the transcription rate. The transcription factor proteins themselves are encoded by genes which are regulated by other transcription factors. These transcription factors might have been regulated by other set of transcription factors in the genome.

Such a complexity leads to a set of interactions which may be visualised to form a transcription network. Thus a transcription network depicts all interactions of the transcription regulatory proteins inside the cell. In such a network the genes are represented by nodes and the edges represents the transcription regulation of one gene by the protein product of the other. Therefore in a transcription network the directed edge X➔Y means that the product of a gene X is a transcription factor protein which controls the rate of transcription of gene Y (by binding to the promoter of gene Y) , which is explained in the following Figure.

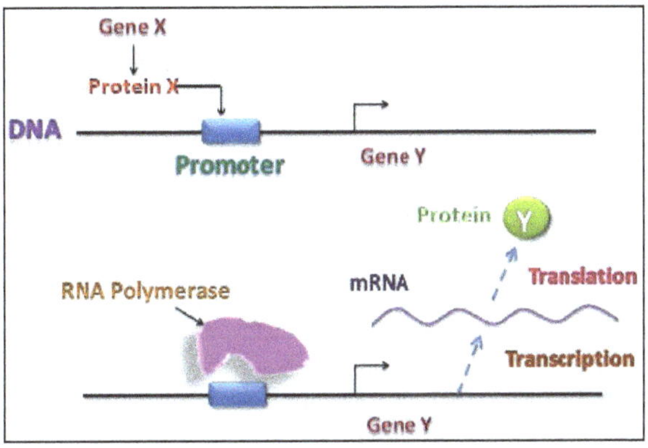

A schematic of transcription and translation of a gene Y.

As shown in the Figure, the network has inputs called signals which represent information from the environment. A signal could be a protein modification, environmental signal, biochemical stimuli, small molecule or a molecular partner that influences one of the transcription factors in the network. The signals initiate a physical change in the transcription factor protein and switch it to an active molecular state. The signal S_X therefore shifts the gene X to its active state X*, binds to the promoter of the gene Y, increasing the rate of transcription and hence increasing the production of protein Y. The gene network is thus a dynamic system. With the arrival of the input signal S_X gene activation profiles change, activities of transcription factors change resulting in changes in protein production rate.

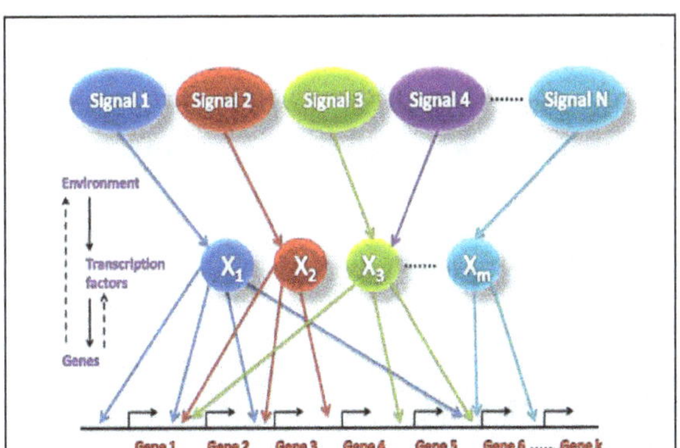

A typical Transcription Network depicting the interaction between the Signals, Transcription Factors and Genes.

Timescales of Transcription Networks

The timescales at which reactions happen in transcription network are quite interesting to note. Table explain timescales of reactions taking place in *E. coli*. We can observe that the input signals such as biochemical stimuli or temperature change the activity of transcription factors on sub second (milliseconds) timescale. The binding of the active transcription factor to its DNA site attains equilibrium in seconds. The transcription and translation of the target gene happens in minutes while the accumulation if protein products and change in the concentration of the stable protein takes hours.

Table: Timescales of reactions in the Transcription Network of the Bacterium *E. coli*	
Binding of a small molecule (a signal) to a transcription factor, causing a change in transcription factor activity.	~1 ms
Binding of active transcription factor to its DNA site.	~1 sec
Transcription + Translation of the gene.	~5 min
Timescale for 50% change in concentration of the translated protein (stable proteins).	~1 h (one cell generation)

Modularity of Transcription Networks

A remarkable property of transcription networks is the modularity of the components of the network. This allows taking off the DNA of a gene from one organism and expressing it on the other. The best example is the isolation of the green fluorescent protein (GFP) from the genome of a jelly fish and to introduce this gene into bacteria causing the bacteria to express green fluorescence.

This modularity confers plasticity to transcription networks during evolution, facilitating introduction of newer genes and newer regulatory pathways. The edges in transcription networks appear to evolve on a faster timescale than the coding regions of the genes. For example, mice and humans are closely related and have similar genes. But their transcription regulation mechanisms and timescale of protein production are quite different.

Signs on the Edges

Each edge in a transcription network corresponds to an interaction in which a transcription factor directly controls the transcription rate of a gene. Such interactions can be classified into two types:

1. Activation or positive control.
2. Repression or negative control.

Activation or Positive Control

When a transcription factor binds to the promoter and increases the rate of transcription it is referred to as activation or positive control. Figure clearly explains the concept further.

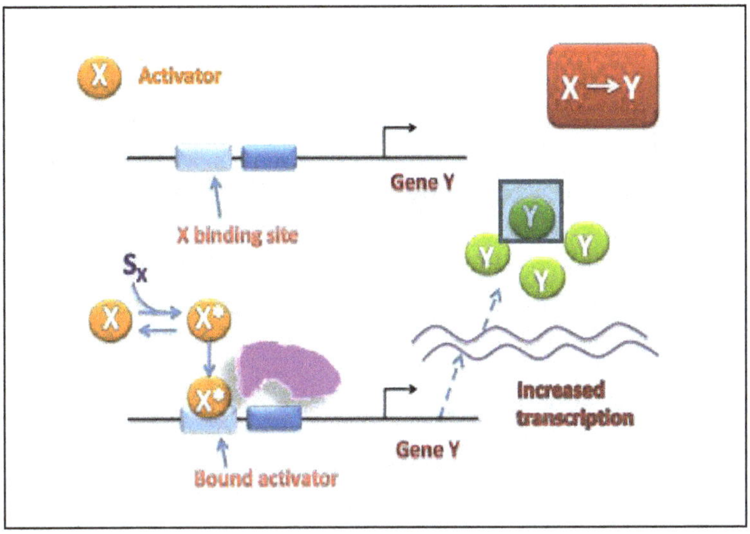

Binding of an activator protein (X*) iIncreases the rate of Transcription of gene Y.

Repression or Negative Control

When a transcription factor binds to the promoter and decreases the rate of transcription it is referred to as repression or negative control. Figure further explains in detail.

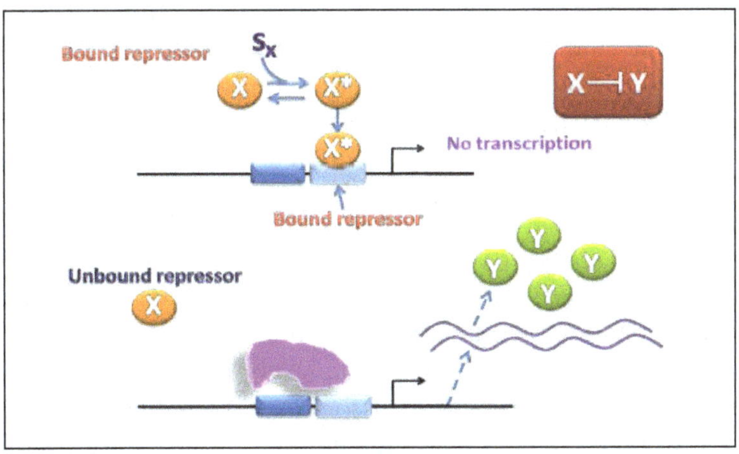

Bound and unbound repressor protein (X*) regulate the rate of transcription of gene Y.

Hence each edge in the network is represented by a plus sign for activation and a minus sign for repression. Generally transcription networks show more positive interactions than negative interactions. For example organisms such as *E. coli* and yeast show 60-80% activation interactions.

Biological Network Inference

Biological network inference is the process of making inferences and predictions about biological networks.

Biological Networks

In a topological sense, a network is a set of nodes and a set of directed or undirected edges between the nodes. Many types of biological networks exist, including transcriptional, signalling and metabolic. Few such networks are known in anything approaching their complete structure, even in the simplest bacteria. Still less is known on the parameters governing the behavior of such networks over time, how the networks at different levels in a cell interact, and how to predict the complete state description of a eukaryotic cell or bacterial organism at a given point in the future. Systems biology, in this sense, is still in its infancy.

There is great interest in network medicine for the modelling biological systems. This article focuses on a necessary prerequisite to dynamic modeling of a network: inference of the topology, that is, prediction of the "wiring diagram" of the network. More specifically, we focus here on inference of biological network structure using the growing sets of high-throughput expression data for genes, proteins, and metabolites. Briefly, methods using high-throughput data for inference of regulatory networks rely on searching for patterns of partial correlation or conditional probabilities that indicate causal influence. Such patterns of partial correlations found in the high-throughput data, possibly combined with other supplemental data on the genes or proteins in the proposed networks, or combined with other information on the organism, form the basis upon which such algorithms work. Such algorithms can be of use in inferring the topology of any network where the change in state of one node can affect the state of other nodes.

Transcriptional Regulatory Networks

Genes are the nodes and the edges are directed. A gene serves as the source of a direct regulatory edge to a target gene by producing an RNA or protein molecule that functions as a transcriptional activator or inhibitor of the target gene. If the gene is an activator, then it is the source of a positive regulatory connection; if an inhibitor, then it is the source of a negative regulatory connection. Computational algorithms take as primary input data measurements of mRNA expression levels of the genes under consideration for inclusion in the network, returning an estimate of the network topology. Such algorithms are typically based on linearity, independence or normality assumptions, which must be verified on a case-by-case basis. Clustering or some form of statistical classification is typically employed to perform an initial organization of the high-throughput mRNA expression values derived from microarray experiments, in particular to select sets of genes as candidates for network nodes. The question then arises: how can the clustering or classification results be connected to the underlying biology? Such results can be useful for pattern classification – for example, to classify subtypes of cancer, or to predict differential responses to a drug (pharmacogenomics). But to understand the relationships between the genes, that is, to more precisely define the influence of each gene on the others, the scientist typically attempts to reconstruct the transcriptional regulatory network. This can be done by data integration in dynamic models supported by background literature, or information in public databases, combined with the clustering results. The modelling can be done by a Boolean network, by Ordinary differential equations or Linear regression models, e.g. Least-angle regression, by Bayesian network or based on Information theory approaches. For instance it can be done by the application of a correlation-based inference algorithm, as will be discussed below, an approach which is having increased success as the size of the available microarray sets keeps increasing

Signal Transduction

Signal transduction networks (very important in the biology of cancer). Proteins are the nodes and directed edges represent interaction in which the biochemical conformation of the child is modified by the action of the parent (e.g. mediated by phosphorylation, ubiquitylation, methylation, etc.). Primary input into the inference algorithm would be data from a set of experiments measuring protein activation / inactivation (e.g., phosphorylation / dephosphorylation) across a set of proteins. Inference for such signalling networks is complicated by the fact that total concentrations of signalling proteins will fluctuate over time due to transcriptional and translational regulation. Such variation can lead to statistical confounding. Accordingly, more sophisticated statistical techniques must be applied to analyse such datasets.

Metabolic

Metabolite networks. Metabolites are the nodes and the edges are directed. Primary input into an algorithm would be data from a set of experiments measuring metabolite levels.

Protein-protein Interaction

Protein-protein interaction networks are also under very active study. However, reconstruction of these networks does not use correlation-based inference in the sense discussed for the networks already described (interaction does not necessarily imply a change in protein state), and a description of such interaction network reconstruction.

Gene co-expression Network

A gene co-expression network (GCN) is an undirected graph, where each node corresponds to a gene, and a pair of nodes is connected with an edge if there is a significant co-expression relationship between them. Having gene expression profiles of a number of genes for several samples or experimental conditions, a gene co-expression network can be constructed by looking for pairs of genes which show a similar expression pattern across samples, since the transcript levels of two co-expressed genes rise and fall together across samples. Gene co-expression networks are of biological interest since co-expressed genes are controlled by the same transcriptional regulatory program, functionally related, or members of the same pathway or protein complex.

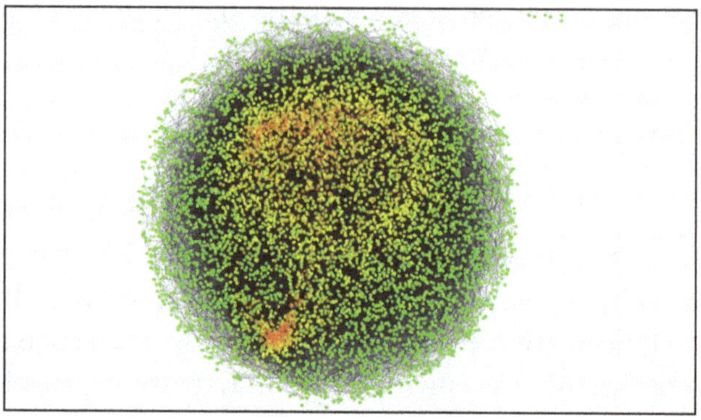

A gene co-expression network constructed from a microarray dataset containing gene expression profiles of 7221 genes for 18 gastric cancer patients.

The direction and type of co-expression relationships are not determined in gene co-expression networks; whereas in a gene regulatory network (GRN) a directed edge connects two genes, representing a biochemical process such as a reaction, transformation, interaction, activation or inhibition. Compared to a GRN, a GCN does not attempt to infer the causality relationships between genes and in a GCN the edges represent only a correlation or dependency relationship among genes. Modules or the highly connected subgraphs in gene co-expression networks correspond to clusters of genes that have a similar function or involve in a common biological process which causes many interactions among themselves.

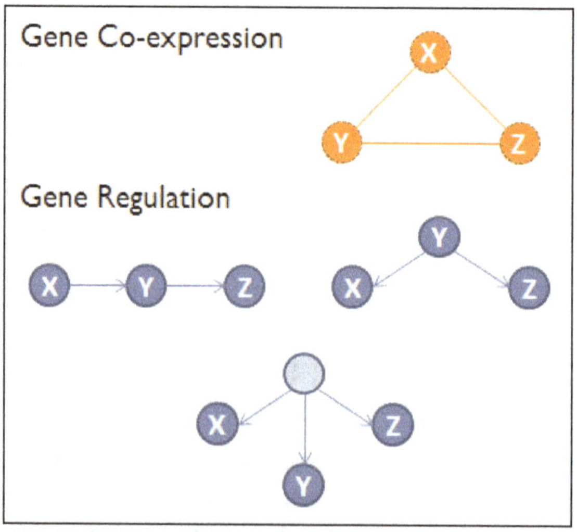

The direction of edges is overlooked in gene co-expression networks. While three genes X, Y and Z are found to be co-expressed, it is not determined whether X activates Y and Y activates Z, or Y activates X and Z, or another gene activates three of them.

Gene co-expression networks are usually constructed using datasets generated by high-throughput gene expression profiling technologies such as Microarray or RNA-Seq.

History

The concept of gene co-expression networks was first introduced by Butte and Kohane in 1999 as *relevance networks*. They gathered the measurement data of medical laboratory tests (e.g. hemoglobin level) for a number of patients and they calculated the Pearson correlation between the results for each pair of tests and the pairs of tests which showed a correlation higher than a certain level were connected in the network (e.g. insulin level with blood sugar). Bute and Kohane used this approach later with mutual information as the co-expression measure and using gene expression data for constructing the first gene co-expression network.

Constructing Gene Co-expression Networks

A good number of methods have been developed for constructing gene co-expression networks. In principle, they all follow a two step approach: calculating co-expression measure, and selecting significance threshold. In first step, a co-expression measure is selected and a similarity score is calculated for each pair of genes using this measure. Then, a threshold is determined and gene pairs which have a similarity score higher than the selected threshold are considered to have significant co-expression relationship and are connected by an edge in the network.

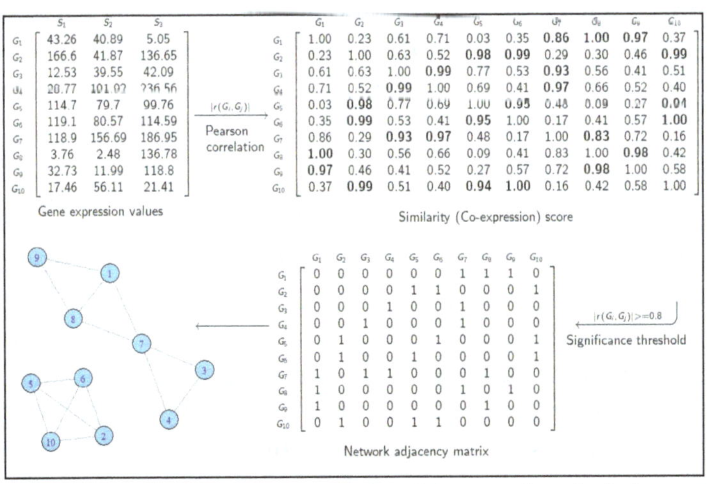

The two general steps for constructing a gene co-expression network: calculating co-expression score (e.g. the absolute value of Pearson correlation coefficient) for each pair of genes, and selecting a significance threshold (e.g. correlation > 0.8).

The input data for constructing a gene co-expression network is often represented as a matrix. If we have the gene expression values of m genes for n samples (conditions), the input data would be an $m \times n$ matrix, called expression matrix. For instance, in a microarray experiment the expression values of thousands of genes are measured for several samples. In first step, a similarity score (co-expression measure) is calculated between each pair of rows in expression matrix. The resulting matrix would be an $m \times m$ matrix, called similarity matrix. Each element in this matrix shows how similar the expression level of two genes change together. In second step, the elements in similarity matrix which are above a certain threshold (i.e. significant co-expressions) are replaced by 1 and the remaining elements are replaced by 0. The resulting matrix, called adjacency matrix, represents the graph of the constructed gene co-expression network. In this matrix, each element shows whether two genes are connected in the network (the 1 elements) or not (the 0 elements).

Co-expression Measure

The expression values of a gene for different samples can be represented as a vector, thus calculating the co-expression measure between a pair of genes is the same as calculating the selected measure for two vectors of numbers.

Pearson's correlation coefficient, Mutual Information, Spearman's rank correlation coefficient and Euclidean distance are the four mostly used co-expression measures for constructing gene co-expression networks. Euclidean distance measures the geometric distance between two vectors, and so considers both the direction and the magnitude of the vectors of gene expression values. Mutual information measures how much knowing the expression levels of one gene reduces the uncertainty about the expression levels of another. Pearson's correlation coefficient measures the tendency of two vectors to increase or decrease together, giving a measure of their overall correspondence. Spearman's rank correlation is the Pearson's correlation calculated for the ranks of gene expression values in a gene expression vector. Several other measures such as partial correlation, regression, and combination of partial correlation and mutual information have also been used.

Each of these measures have their own advantages and disadvantages. The Euclidean distance is not appropriate when the absolute levels of functionally related genes are highly different. Furthermore, if two genes have consistently low expression levels but are otherwise randomly correlated, they might still appear close in Euclidean space. One advantage to mutual information is that it can detect non-linear relationships; however this can turn into a disadvantage because of detecting sophisticated non-linear relationships which does not look biologically meaningful. In addition, for calculating mutual information one should estimate the distribution of the data which needs a large number of samples for a good estimate. Spearman's rank correlation coefficient is more robust to outliers, but on the other hand it is less sensitive to expression values and in datasets with small number of samples may detect many false positives.

Pearson's correlation coefficient is the most popular co-expression measure used in constructing gene co-expression networks. The Pearson's correlation coefficient takes a value between -1 and 1 where absolute values close to 1 show strong correlation. The positive values correspond to an activation mechanism where the expression of one gene increases with the increase in the expression of its co-expressed gene and vice versa. When the expression value of one gene decreases with the increase in the expression of its co-expressed gene, it corresponds to an underlying suppression mechanism and would have a negative correlation.

There are two disadvantages for Pearson correlation measure: it can only detect linear relationships and it is sensitive to outliers. Moreover, Pearson correlation assumes that the gene expression data follow a normal distribution. Song et al. have suggested *biweight midcorrelation (bicor)* as a good alternative for Pearson's correlation. "Bicor is a median based correlation measure, and is more robust than the Pearson correlation but often more powerful than the Spearman's correlation". Furthermore, it has been shown that "most gene pairs satisfy linear or monotonic relationships" which indicates that "mutual information networks can safely be replaced by correlation networks when it comes to measuring co-expression relationships in stationary data".

Threshold Selection

Several methods have been used for selecting a threshold in constructing gene co-expression networks. A simple thresholding method is to choose a co-expression cutoff and select relationships which their co-expression exceeds this cutoff. Another approach is to use Fisher's Z-transformation which calculates a z-score for each correlation based on the number of samples. This z-score is then converted into a p-value for each correlation and a cutoff is set on the p-value. Some methods permute the data and calculate a z-score using the distribution of correlations found between genes in permuted dataset. Some other approaches have also been used such as threshold selection based on clustering coefficient or random matrix theory.

The problem with p-value based methods is that the final cutoff on the p-value is chosen based on statistical routines(e.g. a p-value of 0.01 or 0.05 is considered significant), not based on a biological insight.

WGCNA is a framework for constructing and analyzing weighted gene co-expression networks. The WGCNA method selects the threshold for constructing the network based on the scale-free topology of gene co-expression networks. This method constructs the network for several thresholds and selects the threshold which leads to a network with scale-free topology. Moreover, the WGCNA method constructs a weighted network which means all possible edges appear in the network, but each edge has a weight which shows how significant is the co-expression relationship corresponding to that edge.

References

- Linksvayer, T.; et al. (2012). "Developmental evolution in social insects: Regulatory networks from genes to societies". Journal of Experimental Zoology Part B: Molecular and Developmental Evolution. 318: 159–169. doi:10.1002/jez.b.22001

- Romanuk, T.; et al. (2010). "Maintenance of positive diversity-stability relations along a gradient of environmental stress". PLoS ONE. 5: e10378. doi:10.1371/journal.pone.0010378. PMC 2860506. PMID 20436913
- Faith JJ, Hayete B, Thaden JT, Mogno I, Wierzbowski J, Cottarel G, Kasif S, Collins JJ, Gardner TS (January 2007). "Large-scale mapping and validation of Escherichia coli transcriptional regulation from a compendium of expression profiles". PLoS Biology. 5 (1): e8. PMC 1764438. PMID 17214507. doi:10.1371/journal.pbio.0050008
- Alberts, Bruce; Alexander Johnson; Julian Lewis; Martin Raff; Keith Roberts; Peter Walters (2002). "The Shape and Structure of Proteins". Molecular Biology of the Cell; Fourth Edition. New York and London: Garland Science. ISBN 0-8153-3218-1
- Jambhekar A, Derisi JL (May 2007). "Cis-acting determinants of asymmetric, cytoplasmic RNA transport". RNA. 13 (5): 625–42. doi:10.1261/rna.262607. PMC 1852811. PMID 17449729
- Kober L, Zehe C, Bode J (April 2013). "Optimized signal peptides for the development of high expressing CHO cell lines". Biotechnol. Bioeng. 110 (4): 1164–73. doi:10.1002/bit.24776. PMID 23124363
- Jeremy M. Berg, John L. Tymoczko, Lubert Stryer; Web content by Neil D. Clarke (2002). "3. Protein Structure and Function". Biochemistry. San Francisco: W. H. Freeman. ISBN 0-7167-4684-0
- Amaral PP, Dinger ME, Mercer TR, Mattick JS (March 2008). "The eukaryotic genome as an RNA machine". Science. 319 (5871): 1787–9. Bibcode:2008Sci...319.1787A. doi:10.1126/science.1155472. PMID 18369136
- Berk V, Cate JH (June 2007). "Insights into protein biosynthesis from structures of bacterial ribosomes". Curr. Opin. Struct. Biol. 17 (3): 302–9. doi:10.1016/j.sbi.2007.05.009. PMID 17574829

5

Gene Expression Regulation

The mechanism used by cells to control the increase or decrease in the production of specific gene products is known as the regulation of gene expression. This allows the conservation of both energy and space with the cell. The various control mechanisms and genetic switches are closely examined and explained in an extensive manner in this chapter.

Regulation of gene expression includes a wide range of mechanisms that are used by cells to increase or decrease the production of specific gene products (protein or RNA), and is informally termed *gene regulation*. Sophisticated programs of gene expression are widely observed in biology, for example to trigger developmental pathways, respond to environmental stimuli, or adapt to new food sources. Virtually any step of gene expression can be modulated, from transcriptional initiation, to RNA processing, and to the post-translational modification of a protein. Often, one gene regulator controls another, and so on, in a gene regulatory network.

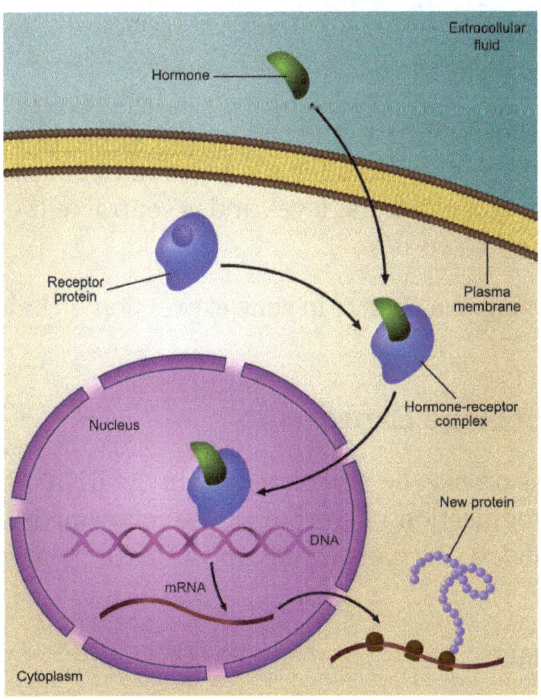

Regulation of gene expression by a hormone receptor.

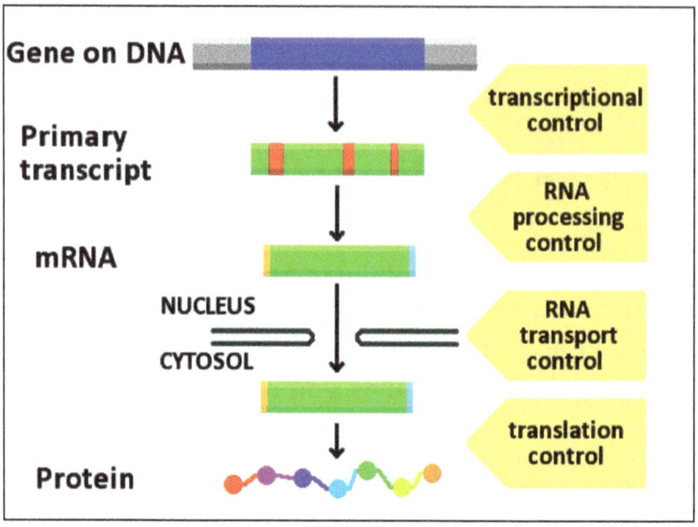

Diagram showing at which stages in the DNA-mRNA-protein pathway expression can be controlled.

Gene regulation is essential for viruses, prokaryotes and eukaryotes as it increases the versatility and adaptability of an organism by allowing the cell to express protein when needed. Although as early as 1951, Barbara McClintock showed interaction between two genetic loci, Activator (*Ac*) and Dissociator (*Ds*), in the color formation of maize seeds, the first discovery of a gene regulation system is widely considered to be the identification in 1961 of the *lac* operon, discovered by Jacques Monod, in which some enzymes involved in lactose metabolism are expressed by *E. coli* only in the presence of lactose and absence of glucose.

In multicellular organisms, gene regulation drives cellular differentiation and morphogenesis in the embryo, leading to the creation of different cell types that possess different gene expression profiles from the same genome sequence. This explains how evolution actually works at a molecular level, and is central to the science of evolutionary developmental biology ("evo-devo").

The initiating event leading to a change in gene expression includes activation or deactivation of receptors.

Regulated Stages of Gene Expression

Any step of gene expression may be modulated, from the DNA-RNA transcription step to post-translational modification of a protein. The following is a list of stages where gene expression is regulated, the most extensively utilised point is Transcription Initiation:

- Chromatin domains.
- Transcription.

- Post-transcriptional modification.
- RNA transport.
- Translation.
- mRNA degradation.

Modification of DNA

In eukaryotes, the accessibility of large regions of DNA can depend on its chromatin structure, which can be altered as a result of histone modifications directed by DNA methylation, ncRNA, or DNA-binding protein. Hence these modifications may up or down regulate the expression of a gene. Some of these modifications that regulate gene expression are inheritable and are referred to as epigenetic regulation.

Structural

Transcription of DNA is dictated by its structure. In general, the density of its packing is indicative of the frequency of transcription. Octameric protein complexes called nucleosomes are responsible for the amount of supercoiling of DNA, and these complexes can be temporarily modified by processes such as phosphorylation or more permanently modified by processes such as methylation. Such modifications are considered to be responsible for more or less permanent changes in gene expression levels.

Chemical

Methylation of DNA is a common method of gene silencing. DNA is typically methylated by methyltransferase enzymes on cytosine nucleotides in a CpG dinucleotide sequence (also called "CpG islands" when densely clustered). Analysis of the pattern of methylation in a given region of DNA (which can be a promoter) can be achieved through a method called bisulfite mapping. Methylated cytosine residues are unchanged by the treatment, whereas unmethylated ones are changed to uracil. The differences are analyzed by DNA sequencing or by methods developed to quantify SNPs, such as Pyrosequencing (Biotage) or MassArray (Sequenom), measuring the relative amounts of C/T at the CG dinucleotide. Abnormal methylation patterns are thought to be involved in oncogenesis.

Histone acetylation is also an important process in transcription. Histone acetyltransferase enzymes (HATs) such as CREB-binding protein also dissociate the DNA from the histone complex, allowing transcription to proceed. Often, DNA methylation and histone deacetylation work together in gene silencing. The combination of the two seems to be a signal for DNA to be packed more densely, lowering gene expression.

Regulation of Transcription

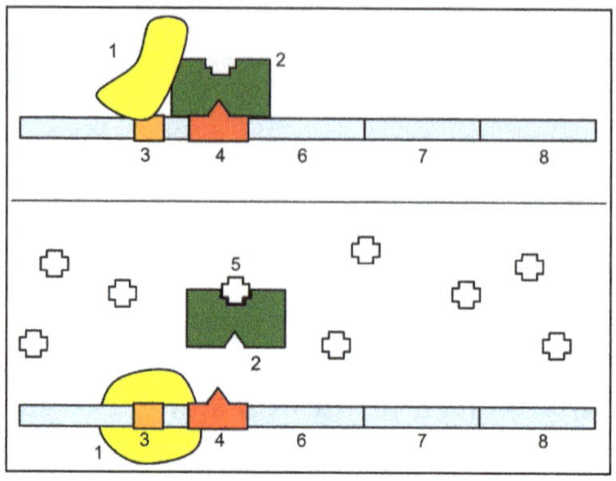

1: RNA Polymerase, *2*: Repressor, *3*: Promoter, *4*: Operator, *5*: Lactose, *6*: lacZ, *7*: lacY, *8*: lacA. Top: The gene is essentially turned off. There is no lactose to inhibit the repressor, so the repressor binds to the operator, which obstructs the RNA polymerase from binding to the promoter and making lactase. Bottom: The gene is turned on. Lactose is inhibiting the repressor, allowing the RNA polymerase to bind with the promoter, and express the genes, which synthesize lactase. Eventually, the lactase will digest all of the lactose, until there is none to bind to the repressor. The repressor will then bind to the operator, stopping the manufacture of lactase.

Regulation of transcription thus controls when transcription occurs and how much RNA is created. Transcription of a gene by RNA polymerase can be regulated by several mechanisms. Specificity factors alter the specificity of RNA polymerase for a given promoter or set of promoters, making it more or less likely to bind to them (i.e., sigma factors used in prokaryotic transcription). Repressors bind to the Operator, coding sequences on the DNA strand that are close to or overlapping the promoter region, impeding RNA polymerase's progress along the strand, thus impeding the expression of the gene.The image to the right demonstrates regulation by a repressor in the lac operon. General transcription factors position RNA polymerase at the start of a protein-coding sequence and then release the polymerase to transcribe the mRNA. Activators enhance the interaction between RNA polymerase and a particular promoter, encouraging the expression of the gene. Activators do this by increasing the attraction of RNA polymerase for the promoter, through interactions with subunits of the RNA polymerase or indirectly by changing the structure of the DNA. Enhancers are sites on the DNA helix that are bound by activators in order to loop the DNA bringing a specific promoter to the initiation complex. Enhancers are much more common in eukaryotes than prokaryotes, where only a few examples exist (to date). Silencers are regions of DNA sequences that, when bound by particular transcription factors, can silence expression of the gene.

Regulation of Transcription in Cancer

In vertebrates, the majority of gene promoters contain a CpG island with numerous CpG sites. When many of a gene's promoter CpG sites are methylated the gene becomes silenced. Colorectal cancers typically have 3 to 6 driver mutations and 33 to 66 hitchhiker or passenger mutations. However, transcriptional silencing may be of more importance than mutation in causing progression to cancer. For example, in colorectal cancers about 600 to 800 genes are transcriptionally silenced by CpG island methylation. Transcriptional repression in cancer can also occur by other epigenetic mechanisms, such as altered expression of microRNAs. In breast cancer, transcriptional repression of BRCA1 may occur more frequently by over-expressed microRNA-182 than by hypermethylation of the BRCA1 promoter.

Post-transcriptional Regulation

After the DNA is transcribed and mRNA is formed, there must be some sort of regulation on how much the mRNA is translated into proteins. Cells do this by modulating the capping, splicing, addition of a Poly(A) Tail, the sequence-specific nuclear export rates, and, in several contexts, sequestration of the RNA transcript. These processes occur in eukaryotes but not in prokaryotes. This modulation is a result of a protein or transcript that, in turn, is regulated and may have an affinity for certain sequences.

Three Prime Untranslated Regions and MicroRNAs

Three prime untranslated regions (3'-UTRs) of messenger RNAs (mRNAs) often contain regulatory sequences that post-transcriptionally influence gene expression. Such 3'-UTRs often contain both binding sites for microRNAs (miRNAs) as well as for regulatory proteins. By binding to specific sites within the 3'-UTR, miRNAs can decrease gene expression of various mRNAs by either inhibiting translation or directly causing degradation of the transcript. The 3'-UTR also may have silencer regions that bind repressor proteins that inhibit the expression of a mRNA.

The 3'-UTR often contains miRNA response elements (MREs). MREs are sequences to which miRNAs bind. These are prevalent motifs within 3'-UTRs. Among all regulatory motifs within the 3'-UTRs (e.g. including silencer regions), MREs make up about half of the motifs.

As of 2014, the miRBase web site, an archive of miRNA sequences and annotations, listed 28,645 entries in 233 biologic species. Of these, 1,881 miRNAs were in annotated human miRNA loci. miRNAs were predicted to have an average of about four hundred target mRNAs (affecting expression of several hundred genes). Freidman et al. estimate that >45,000 miRNA target sites within human mRNA 3'-UTRs are conserved above background levels, and >60% of human protein-coding genes have been under selective pressure to maintain pairing to miRNAs.

Direct experiments show that a single miRNA can reduce the stability of hundreds of unique mRNAs. Other experiments show that a single miRNA may repress the production of hundreds of proteins, but that this repression often is relatively mild (less than 2-fold).

The effects of miRNA dysregulation of gene expression seem to be important in cancer. For instance, in gastrointestinal cancers, a 2015 paper identified nine miRNAs as epigenetically altered and effective in down-regulating DNA repair enzymes.

The effects of miRNA dysregulation of gene expression also seem to be important in neuropsychiatric disorders, such as schizophrenia, bipolar disorder, major depressive disorder, Parkinson's disease, Alzheimer's disease and autism spectrum disorders.

Regulation of Translation

The translation of mRNA can also be controlled by a number of mechanisms, mostly at the level of initiation. Recruitment of the small ribosomal subunit can indeed be modulated by mRNA secondary structure, antisense RNA binding, or protein binding. In both prokaryotes and eukaryotes, a large number of RNA binding proteins exist, which often are directed to their target sequence by the secondary structure of the transcript, which may change depending on certain conditions, such as temperature or presence of a ligand (aptamer). Some transcripts act as ribozymes and self-regulate their expression.

Examples of Gene Regulation

- Enzyme induction is a process in which a molecule (e.g., a drug) induces (i.e., initiates or enhances) the expression of an enzyme.
- The induction of heat shock proteins in the fruit fly *Drosophila melanogaster*.
- The Lac operon is an interesting example of how gene expression can be regulated.
- Viruses, despite having only a few genes, possess mechanisms to regulate their gene expression, typically into an early and late phase, using collinear systems regulated by anti-terminators (lambda phage) or splicing modulators (HIV).
- GAL4 is a transcriptional activator that controls the expression of GAL1, GAL7, and GAL10 (all of which code for the metabolic of galactose in yeast). The GAL4/UAS system has been used in a variety of organisms across various phyla to study gene expression.

Developmental Biology

A large number of studied regulatory systems come from developmental biology. Examples include:

- The colinearity of the Hox gene cluster with their nested antero-posterior patterning.

- It has been speculated that pattern generation of the hand (digits - interdigits) The gradient of Sonic hedgehog (secreted inducing factor) from the zone of polarizing activity in the limb, which creates a gradient of active Gli3, which activates Gremlin, which inhibits BMPs also secreted in the limb, resulting in the formation of an alternating pattern of activity as a result of this reaction-diffusion system.
- Somitogenesis is the creation of segments (somites) from a uniform tissue (Pre-somitic Mesoderm). They are formed sequentially from anterior to posterior. This is achieved in amniotes possibly by means of two opposing gradients, Retinoic acid in the anterior (wavefront) and Wnt and Fgf in the posterior, coupled to an oscillating pattern (segmentation clock) composed of FGF + Notch and Wnt in antiphase.
- Sex determination in the soma of a Drosophila requires the sensing of the ratio of autosomal genes to sex chromosome-encoded genes, which results in the production of sexless splicing factor in females, resulting in the female isoform of doublesex.

Circuitry

Up-regulation and Down-regulation

Up-regulation is a process that occurs within a cell triggered by a signal (originating internal or external to the cell), which results in increased expression of one or more genes and as a result the protein(s) encoded by those genes. Conversely, down-regulation is a process resulting in decreased gene and corresponding protein expression.

- Up-regulation occurs, for example, when a cell is deficient in some kind of receptor. In this case, more receptor protein is synthesized and transported to the membrane of the cell and, thus, the sensitivity of the cell is brought back to normal, reestablishing homeostasis.
- Down-regulation occurs, for example, when a cell is overstimulated by a neurotransmitter, hormone, or drug for a prolonged period of time, and the expression of the receptor protein is decreased in order to protect the cell.

Inducible vs. Repressible Systems

Gene Regulation can be summarized by the response of the respective system:

- Inducible systems - An inducible system is off unless there is the presence of some molecule (called an inducer) that allows for gene expression. The molecule is said to “induce expression”. The manner by which this happens is dependent on the control mechanisms as well as differences between prokaryotic and eukaryotic cells.

- Repressible systems - A repressible system is on except in the presence of some molecule (called a corepressor) that suppresses gene expression. The molecule is said to "repress expression". The manner by which this happens is dependent on the control mechanisms as well as differences between prokaryotic and eukaryotic cells.

The GAL4/UAS system is an example of both an inducible and repressible system. GAL4 binds an upstream activation sequence (UAS) to activate the transcription of the GAL1/GAL7/GAL10 cassette. On the other hand, a MIG1 response to the presence of glucose can inhibit GAL4 and therefore stop the expression of the GAL1/GAL7/GAL10 cassette.

Theoretical Circuits

- Repressor/Inducer: an activation of a sensor results in the change of expression of a gene.
- negative feedback: the gene product downregulates its own production directly or indirectly, which can result in:
    - Keeping transcript levels constant/proportional to a factor.
    - Inhibition of run-away reactions when coupled with a positive feedback loop.
    - Creating an oscillator by taking advantage in the time delay of transcription and translation, given that the mRNA and protein half-life is shorter.
- positive feedback: the gene product upregulates its own production directly or indirectly, which can result in:
    - Signal amplification.
    - Bistable switches when two genes inhibit each other and both have positive feedback.
    - pattern generation.

Study Methods

In general, most experiments investigating differential expression used whole cell extracts of RNA, called steady-state levels, to determine which genes changed and by how much they did. These are, however, not informative of where the regulation has occurred and may actually mask conflicting regulatory processes, but it is still the most commonly analysed (quantitative PCR and DNA microarray).

When studying gene expression, there are several methods to look at the various stages. In eukaryotes these include:

- The local chromatin environment of the region can be determined by ChIP-chip analysis by pulling down RNA Polymerase II, Histone 3 modifications, Trithorax-group protein, Polycomb-group protein, or any other DNA-binding element to which a good antibody is available.
- Epistatic interactions can be investigated by synthetic genetic array analysis
- Due to post-transcriptional regulation, transcription rates and total RNA levels differ significantly. To measure the transcription rates nuclear run-on assays can be done and newer high-throughput methods are being developed, using thiol labelling instead of radioactivity.
- Only 5% of the RNA polymerised in the nucleus actually exits, and not only introns, abortive products, and non-sense transcripts are degradated. Therefore, the differences in nuclear and cytoplasmic levels can be see by separating the two fractions by gentle lysis.
- Alternative splicing can be analysed with a splicing array or with a tiling array.
- All in vivo RNA is complexed as RNPs. The quantity of transcripts bound to specific protein can be also analysed by RIP-Chip. For example, DCP2 will give an indication of sequestered protein; ribosome-bound gives and indication of transcripts active in transcription (although it should be noted that a more dated method, called polysome fractionation, is still popular in some labs).
- Protein levels can be analysed by Mass spectrometry, which can be compared only to quantitative PCR data, as microarray data is relative and not absolute.
- RNA and protein degradation rates are measured by means of transcription inhibitors (actinomycin D or α-amanitin) or translation inhibitors (Cycloheximide), respectively.

Need for Gene Regulation

The biochemical processes and signalling inside cells is mostly enzyme driven and are energy dependent. The enzymatic proteins which affect these processes are products of gene transcription and translation. These genes are switched on or off depend on the requirements of the time scales.

Gene switching happens due to a variety of reasons, some of which are listed below:

- Selective switching of genes prevents wastage of energy and cellular resources. Many genes posses the natural capability to switch ON or switch OFF genes based on their requirements.

- Gene switching events are one of the key processes required at development phases of multi cellular organisms during cell fate decisions.
- Unregulated gene expression leads to diseases for example, the onset of cancer can be attributed to aberrant gene expression pattern where cancer cells lose their ability to regulate mitosis-an uncontrolled cell division.

How do External Signals Control Gene Expression?

Extra cellular cues from the environment like temperature, changes in pH and changes in internal biochemical signalling have been shown to alter gene expression to a significant extent. The response of a liver cell to the steroid- gluoco corticoid hormone is a striking example of altered gene expression. Gluoco corticoid hormones are released in the body during starvation or during extreme activity of exercising. Gluoco corticoid signalling in liver enhances glucose production and alters expression of tyrosine amino transferase which aids the conversion from tyrosine to glucose. In the absence of the hormone, basal levels of these enzymes are maintained at basic levels. Adipose cells respond differently to gluoco corticoids. When a fat cell is exposed to gluoco corticoid, the levels of tyrosine amino transferases are reduced. Some of the cell types do not respond to gluoco corticoid at all. We therefore understand that the same extracellular signal elicits different responses from different cell types.

Gene Switching

A differentiated cell example neuronal cell performs its unique and specific function by switching ON a specific set of genes and by switching OFF another. Some of these genes are permanently active (switched ON) because they carry out vital functions of the cell and because they contain the blueprint for vital enzymes required for cellular function. Certain genes not required for a particular function in the cell may be switched OFF. The production of insulin in pancreatic cell signals the ON state of the insulin gene while other genes irrelevant for insulin signalling may be switched OFF.

The level of expression of a specific gene in a cell is a major factor in the expression control mechanism. For example, the amount of melanin expressed by the skin cell decides the color of the skin. Melanin production is under the type control of two genes each of which show dominant and recessive expression as represented in Figure.

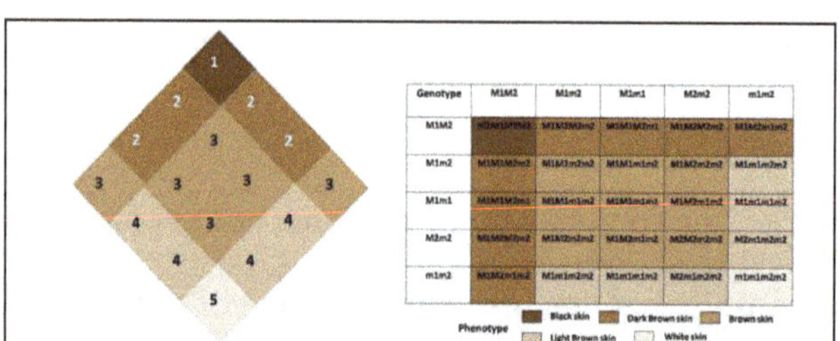

Genotypes and phenotypes expressed by 2 melanin genes (dominant and recessive).

The Figure illustrates 5 possible genotypes that the melanin gene can express. We find that 4 dominant alleles represent enhanced expression of melanin leading to the phenotype of black skin.

Gene Control in Eukaryotic Systems

Studying gene control in eukaryotic cells is an active and attractive pursuit for experimental biologists. The term 'Eukaryotic Gene Control' is not restricted to the process of transcribing to mRNA. For example, the transcription units for mRNAs do not mark the final mRNA product in the case of vertebrate cells. Several control points exists from the processing of primary RNA transcripts in the cell nucleus till the mRNA is translated in the cytoplasm. Specifically the eukaryotic genes undergo more levels of regulation than the bacterial genes. Transcription regulation, RNA processing in the nucleus, mRNA stability, and the frequency of translation events in the cytoplasm are the vital steps at which eukaryotic genes exercise control in expression. The same has been illustrated in Figure.

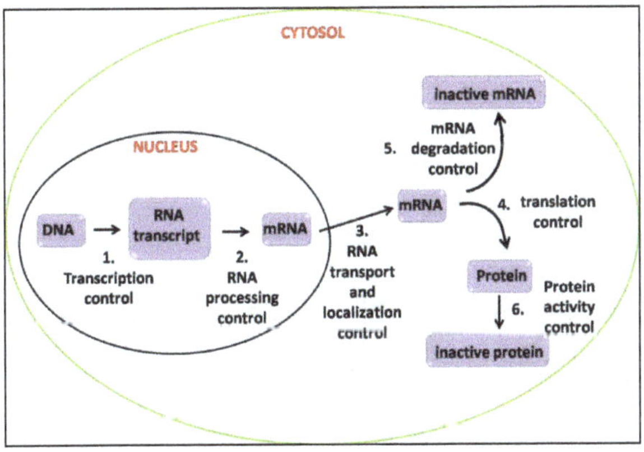

Gene regulatory pathways extending from the nucleus to the cytoplasm of a typical eukaryotic cell.

The transcription process in eukaryotic cells is more complex than in the prokaryotic systems. Gene regulation in these systems can be effected in two ways as shown in the Figure.

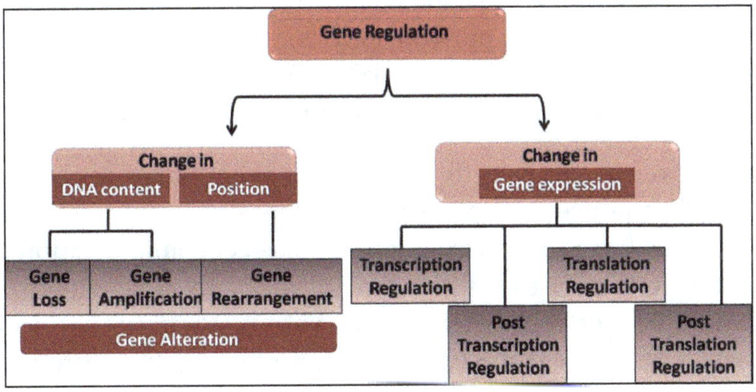

Mechanisms of gene regulation in eukaryotes.

The regulation could be effected through:

1 Control mechanisms that involve change in DNA content or position. Here the DNA of somatic cells can be altered but these changes do not affect gametes so that they are not inherited by the offspring.

2 Control mechanisms involving changes in expression patterns. Regulatory mechanisms involving change in DNA content or position involves alterations at the level of the gene and may be due to:

(a) Gene loss.

(b) Gene amplification.

(c) Gene rearrangement.

Regulatory mechanisms involving changes in expression pattern include:

(a) Transcriptional control.

(b) Post-transcriptional control.

(c) Translational control.

(d) Post-translational control.

Gene alteration is one of the key mechanisms of gene regulation in eukaryotes. This may be carried out through mechanisms like gene loss, gene amplification or through gene rearrangement. Gene alteration plays a role in regulating gene expression under developmental stages, drug resistance etc. Gene loss involves elimination of the gene from differentiating cells in protozoans as well as in metazoans. Lineage specific gene loss is a major evolutionary process studied from gene sets of completely sequenced genomes. In prokaryotes, gene loss and horizontal gene transfer contributes to an intensive gene flux that shapes the genomes of these organisms. In eukaryotes lineage specific gene loss is greater than in prokaryotes. A very strong selection pressures result in dramatic gene loss. For example, Encephalitozoon cuniculi is a eukaryotic intracellular parasite, which now has been estimated to have approximately 2000 genes compared to around 6000 genes in the genomes of yeast, representing a striking case of massive gene loss. Complete chromosomes are preserved only in cells serving as precursors for gamete formation.

Gene amplification refers to a selective increase in the number of copies of a gene sequence. Cancer cells produce multiple copies of genes due to extra cellular signalling or in response to signals from the environment. Hence gene amplification occurs during malignancy and in certain cases cells become resistant to drugs due to the process of gene amplification. Cancer cells when treated with methotrexate which blocks dihydrofolate reductase undergo gene amplification producing dihydrofolate reductase in abundant amount to circumvent the effect of methotrexate. The term 'gene amplification' is often used to refer to the laboratory technique called 'polymerase Chain Reaction' which enables amplifying gene sequences in an eppendorf tube.

Gene rearrangement involves the translocation of a gene from one region of a genome to another. This involves rearrangement of substantial stretches of DNA in the somatic cells and can alter gene expression. Transposons are DNA stretches that move across locations within the genome. These transposons control gene expression depending on their location. For example, if the transposon is inserted in a transcription regulatory sequence it regulates the levels of protein production. It may also carry a gene that is activated if it is inserted downstream of a promoter sequence. Mammalian systems such as mouse or human being synthesize large amounts of specific antibodies against any foreign antigens on exposure. Antibodies are composed of two heavy and two light chains. The specificity of the antibody is determined by the amino acid sequences of the variable regions of both the light and the heavy chains. The rearrangement of various immunoglobulin genes leads to the formation of a variety of antibodies specific for diverse antigens. African Trypanosomes sequentially express various surface antigens in their mammalian host. The expression of such Variable Surface Glycoprotein (VSG) antigens has been shown to be related to DNA rearrangements. (Williams R. et al., Nature 1979, Payes et al. PNAS 1981). Trypanosomes circumvent host defense by replacing a copy of VSG (Variant Surface Glycoprotein) gene at the expression site by another VSG gene at some other site.

We shall discuss in the next class how changes in gene expression can be brought about at the level of transcription, translation and at the level of post transcription and post translation through precise regulatory mechanisms that control expressions.

Gene Control Mechanisms at Transcriptional and Post Transcriptional Levels

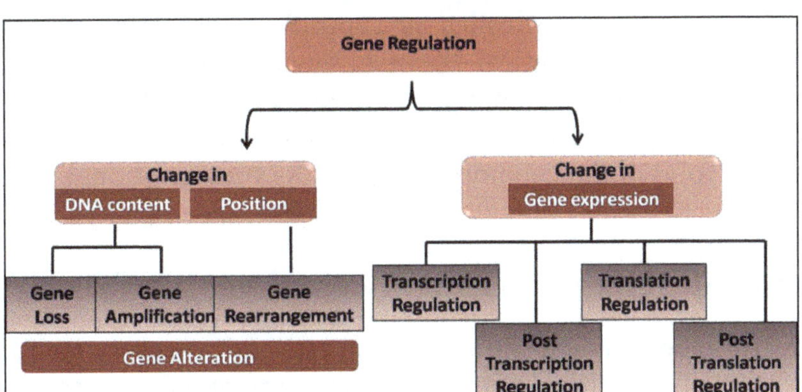

Mechanisms of gene regulation in eukaryotes.

The complicated mechanisms of mRNA formation, the complex structures of the nucleus and cytoplasm of eukaryotic cells, the robust genetic program involved in differentiation of eukaryotic cells strikingly differentiate eukaryotic gene control from bacterial gene regulation. As stated earlier, the eukaryotic genes undergo several more steps of regulation than the bacterial genes. Single molecule experiments have clearly and accurately decipher the mechanical steps involved in eukaryotic gene control. Such experiments have even determined at the transcription level, the amount of specific mRNA molecules that are regulated, the processing of the nuclear RNA and the stability of mRNA. As stated in the previous class, gene regulation can be brought about by altering gene expression states in a cell. This can be effected at the levels of transcription and translation and during post transcription and post translational events.

Transcriptional Regulation

In molecular biology and genetics, transcriptional regulation is the means by which a cell regulates the conversion of DNA to RNA (transcription), thereby orchestrating gene activity. A single gene can be regulated in a range of ways, from altering the number of copies of RNA that are transcribed, to the temporal control of when the gene is transcribed. This control allows the cell or organism to respond to a variety of intra- and extracellular signals and thus mount a response. Some examples of this include producing the mRNA that encode enzymes to adapt to a change in a food source, producing the gene products involved in cell cycle specific activities, and producing the gene products responsible for cellular differentiation in higher eukaryotes, as studied in evolutionary developmental biology.

The regulation of transcription is a vital process in all living organisms. It is orchestrated by transcription factors and other proteins working in concert to finely tune the amount of RNA being produced through a variety of mechanisms. Prokaryotic organisms and eukaryotic organisms have very different strategies of accomplishing control over transcription, but some important features remain conserved between the two. Most importantly is the idea of combinatorial control, which is that any given gene is likely controlled by a specific combination of factors to control transcription. In a hypothetical example, the factors A and B might regulate a distinct set of genes from the combination of factors A and C. This combinatorial nature extends to complexes of far more than two proteins, and allows a very small subset (less than 10%) of the genome to control the transcriptional program of the entire cell.

Transcription Regulation in Prokaryotes

Much of the early understanding of transcription came from prokaryotic organisms, although the extent and complexity of transcriptional regulation is greater in eukaryotes. Prokaryotic transcription is governed by three main sequence elements:

- Promoters are elements of DNA that may bind RNA polymerase and other proteins for the successful initiation of transcription directly upstream of the gene.
- Operators recognize repressor proteins that bind to a stretch of DNA and inhibit the transcription of the gene.
- Positive control elements that bind to DNA and incite higher levels of transcription.

While these means of transcriptional regulation also exist in eukaryotes, the transcriptional landscape is significantly more complicated both by the number of proteins involved as well as by the presence of introns and the packaging of DNA into histones.

The transcription of a basic prokaryotic gene is dependent on the strength of its promoter and the presence of activators or repressors. In the absence of other regulatory elements, a promoter's sequence-based affinity for RNA polymerases varies, which results in the production of different amounts of transcript. The variable affinity of RNA polymerase for different promoter sequences is related to regions of consensus sequence upstream of the transcription start site. The more nucleotides of a promoter that agree with the consensus sequence, the stronger the affinity of the promoter for RNA Polymerase likely is.

In the absence of other regulatory elements, the default state of a prokaryotic transcript is to be in the "on" configuration, resulting in the production of some amount of transcript. This means that transcriptional regulation in the form of protein repressors and positive control elements can either increase or decrease transcription. Repressors often physically occupy the promoter location, occluding RNA polymerase from binding. Alternatively a repressor and polymerase may bind to the DNA at the same time with a physical interaction between the repressor preventing the opening of the DNA for access to the minus strand for transcription. This strategy of control is distinct from eukaryotic transcription, whose basal state is to be off and where co-factors required for transcription initiation are highly gene dependent.

Sigma factors are specialized bacterial proteins that bind to RNA polymerases and orchestrate transcription initiation. Sigma factors act as mediators of sequence-specific transcription, such that a single sigma factor can be used for transcription of all housekeeping genes or a suite of genes the cell wishes to express in response to some external stimuli such as stress.

Eukaryotic Transcription Regulation

The added complexity of generating a eukaryotic cell carries with it an increase in the complexity of transcriptional regulation. Eukaryotes have three RNA polymerases, known as Pol I, Pol II, and Pol III. Each polymerase has specific targets and activities, and is regulated by independent mechanisms. There are a number of additional mechanisms through which polymerase activity can be controlled. These mechanisms can be generally grouped into three main areas:

- Control over polymerase access to the gene. This is perhaps the broadest of the three control mechanisms. This includes the functions of histone remodeling enzymes, transcription factors, enhancers and repressors, and many other complexes.
- Productive elongation of the RNA transcript. Once polymerase is bound to a promoter, it requires another set of factors to allow it to escape the promoter complex and begin successfully transcribing RNA.
- Termination of the polymerase. A number of factors which have been found to control how and when termination occurs, which will dictate the fate of the RNA transcript.

All three of these systems work in concert to integrate signals from the cell and change the transcriptional program accordingly.

While in prokaryotic systems the basal transcription state can be thought of as nonrestrictive (that is, "on" in the absence of modifying factors), eukaryotes have a restrictive basal state which requires the recruitment of other factors in order to generate RNA transcripts. This difference is largely due to the compaction of the eukaryotic genome by winding DNA around histones to form higher order structures. This compaction makes the gene promoter inaccessible without the assistance of other factors in the nucleus, and thus chromatin structure is a common site of regulation. Similar to the sigma factors in prokaryotes, the general transcription factors (GTFs) are a set of factors in eukaryotes that are required for all transcription events. These factors are responsible for stabilizing binding interactions and opening the DNA helix to allow the RNA polymerase to access the template, but generally lack specificity for different promoter sites. A large part of gene regulation occurs through transcription factors that either recruit or inhibit the binding of the general transcription machinery and/or the polymerase. This can be accomplished through close interactions with core promoter elements, or through the long distance enhancer elements.

Once a polymerase is successfully bound to a DNA template, it often requires the assistance of other proteins in order to leave the stable promoter complex and begin elongating the nascent RNA strand. This process is called promoter escape, and is another step at which regulatory elements can act to accelerate or slow the transcription process. Similarly, protein and nucleic acid factors can associate with the elongation complex and modulate the rate at which the polymerase moves along the DNA template.

Regulation at the Level of Chromatin State

In eukaryotes, genomic DNA is highly compacted in order to be able to fit it into the nucleus. This is accomplished by winding the DNA around protein octamers called histones, which has consequences for the physical accessibility of parts of the genome at any given time. Significant portions are silenced through histone modifications, and thus are inaccessible to the polymerases or their cofactors. The highest level of transcription regulation occurs through the rearrangement of histones in order to expose or sequester genes, because these processes have the ability to render entire regions of a chromosome inaccessible such as what occurs in imprinting.

Histone rearrangement is facilitated by post-translational modifications to the tails of the core histones. A wide variety of modifications can be made by enzymes such as the histone acetyltransferases (HATs), histone methyltransferases (HMTs), and histone deacetylases (HDACs), among others. These enzymes can add or remove covalent modifications such as methyl groups, acetyl groups, phosphates, and ubiquitin. Histone modifications serve to recruit other proteins which can either increase the compaction of the chromatin and sequester promoter elements, or to increase the spacing between histones and allow the association of transcription factors or polymerase on open DNA. For example, H3K27 trimethylation by the polycomb complex PRC2 causes chromosomal compaction and gene silencing. These histone modifications may be created by the cell, or inherited in an epigenetic fashion from a parent.

Regulation through Transcription Factors and Enhancers

Transcription Factors

Transcription factors are proteins that bind to specific DNA sequences in order to regulate the expression of a given gene. The power of transcription factors resides in their ability to activate and/or repress wide repertoires of downstream target genes. The fact that these transcription factors work in a combinatorial fashion means that only a small subset of an organism's genome encodes transcription factors. Transcription factors function through a wide variety of mechanisms. Often they are at the end of a signal transduction pathway that functions to change something about the factor, like its subcellular localization or its activity. Post-translational modifications to transcription factors located in the cytosol can cause them to translocate to the nucleus where they can interact with their corresponding enhancers. Others are already in the nucleus, and are modified to enable the interaction with partner transcription factors. Some post-translational modifications known to regulate the functional state of transcription factors are phosphorylation, acetylation, SUMOylation and ubiquitylation. Transcription factors can be divided in two main categories: activators and repressors. While activators can interact directly or indirectly with the core machinery of transcription through enhancer binding, repressors predominantly recruit co-repressor complexes leading to transcriptional repression by chromatin condensation of enhancer regions. It may also happen that a repressor may function by allosteric competition against a determined activator to repress gene expression: overlapping DNA-binding motifs for both activators and repressors induce a physical competition to occupy the site of binding. If the repressor has a higher affinity for its motif than the activator, transcription would be effectively blocked in the presence of the repressor. Tight regulatory control is achieved by the highly dynamic nature of transcription factors. Again, many different mechanisms exist to control whether a transcription factor is active. These mechanisms include control over protein localization or control over whether the protein can bind DNA. An example of this is the protein HSF1, which remains bound to Hsp70 in the cytosol and is only translocated into the nucleus upon cellular stress such as heat shock. Thus the genes under the control of this transcription factor will remain untranscribed unless the cell is subjected to stress.

Enhancers

Enhancers or cis-regulatory modules/elements (CRM/CRE) are non-coding DNA sequences containing multiple activator and repressor binding sites. Enhancers range from 200 bp to 1 kb in length and can be either proximal, 5' upstream to the promoter or within the first intron of the regulated gene, or distal, in introns of neighboring genes or intergenic regions far away from the locus. Through DNA looping, active enhancers contact the promoter dependently of the core DNA binding motif promoter specificity. Promoter-enhancer dichotomy provides the basis for the functional interaction between transcription factors and transcriptional core machinery to trigger RNA Pol II escape from the promoter. Whereas one could think that there is a 1:1 enhancer-promoter ratio, studies of the human genome predict that an active promoter interacts with 4 to 5 enhancers. Similarly, enhancers can regulate more than one gene without linkage restriction and are said to "skip" neighboring genes to regulate more distant ones. Even though infrequent, transcriptional regulation can involve elements located in a chromosome different to one where the promoter resides. More interestingly, proximal enhancers or promoters of neighboring genes can serve as platforms to recruit more distal elements.

Regulatory Landscape

Transcriptional initiation, termination and regulation are mediated by "DNA looping" which brings together promoters, enhancers, transcription factors and RNA processing factors to accurately regulate gene expression. Chromosome conformation capture (3C) and more recently Hi-C techniques provided evidence that active chromatin regions are "compacted" in nuclear domains or bodies where transcriptional regulation is enhanced24614317. The configuration of the genome is essential for enhancer-promoter proximity. Cell-fate decisions are mediated upon highly dynamic genomic reorganizations at interphase to modularly switch on or off entire gene regulatory networks through short to long range chromatin rearrangements. Related studies demonstrate that metazoan genomes are partitioned in structural and functional units around a megabase long called Topological association domains (TADs) containing dozens of genes regulated by hundreds of enhancers distributed within large genomic regions containing only non-coding sequences. The function of TADs is to regroup enhancers and promoters interacting together within a single large functional domain instead of having them spread in different TADs. However, studies of mouse development point out that two adjacent TADs may regulate the same gene cluster. The most relevant study on limb evolution shows that the TAD at the 5' of the HoxD gene cluster in tetrapod genomes drives its expression in the distal limb bud embryos, giving rise to the hand, while the one located at 3' side does it in the proximal limb bud, giving rise to the arm. Still, it is not known whether TADs are an adaptive strategy to enhance regulatory interactions or an effect of the constrains on these same interactions. TAD boundaries are often composed by housekeeping genes, tRNAs, other highly expressed sequences and Short Interspersed Elements (SINE). While these genes may take advantage of their border position to be ubiquitously expressed, they are not directly linked with TAD edge formation. The specific molecules identified at boundaries of TADs are called insulators or architectural proteins because they not only block enhancer leaky expression but also ensure an accurate compartmentalization of cis-regulatory inputs to the targeted promoter. These insulators are DNA-binding proteins like CTCF and TFIIIC that help recruiting structural partners such as cohesins and condensins. The localization and binding of architectural proteins to their corresponding binding sites is regulated by post-translational modifications. Interestingly, DNA binding motifs recognized by architectural proteins are either of high occupancy and at around a megabase of each other or of low occupancy and inside TADs. High occupancy sites are usually conserved and static while intra-TADs sites are dynamic according to the state of the cell therefore TADs themselves are compartmentalized in subdomains that can be called subTADs from few kb up to a TAD long (19). When architectural binding sites are at less than 100 kb from each other, Mediator proteins are the architectural proteins cooperate with cohesin. For subTADs larger than 100 kb and TAD boundaries, CTCF is the typical insulator found to interact with cohesion.

Regulation of the Pre-initiation Complex and Promoter Escape

In eukaryotes, ribosomal rRNA and the tRNAs involved in translation are controlled by RNA polymerase I (Pol I) and RNA polymerase III (Pol III). RNA Polymerase II is responsible for the production of messenger RNA (mRNA) within the cell. Particularly for Pol II, much of the regulatory checkpoints in the transcription process occur in the assembly and escape of the pre-initiation complex. A gene-specific combination of transcription factors will recruit TFIID and/or TFIIA to the core promoter, followed by the association of TFIIB, creating a stable complex onto which the rest of the General Transcription Factors (GTFs) can assemble. This complex is relatively stable, and can undergo multiple rounds of transcription initiation. After the binding of TFIIB and TFIID, Pol II the rest of the GTFs can assemble. This assembly is marked by the post-translational modification (typically phosphorylation) of the C-terminal domain (CTD) of Pol II through a number of kinases. The CTD is a large, unstructured domain extending from the RbpI subunit of Pol II, and consists of many repeats of the heptad sequence YSPTSPS. TFIIH, the helicase that remains associated with Pol II throughout transcription, also contains a subunit with kinase activity which will phosphorylate the serines 5 in the heptad sequence. Similarly, both CDK8 (a subunit of the massive multiprotein Mediator complex) and CDK9 (a subunit of the p-TEFb elongation factor), have kinase activity towards other residues on the CTD. These phosphorylation events promote the transcription process and serve as sites of recruitment for mRNA processing machinery. All three of these kinases respond to upstream signals, and failure to phosphorylate the CTD can lead to a stalled polymerase at the promoter.

Regulation of Transcription in Cancer

In vertebrates, the majority of gene promoters contain a CpG island with numerous CpG sites. When many of a gene's promoter CpG sites are methylated the gene becomes silenced. Colorectal cancers typically have 3 to 6 driver mutations and 33 to 66 hitchhiker or passenger mutations. However, transcriptional silencing may be of more importance than mutation in causing progression to cancer. For example, in colorectal cancers about 600 to 800 genes are transcriptionally silenced by CpG island methylation. Transcriptional repression in cancer can also occur by other epigenetic mechanisms, such as altered expression of microRNAs. In breast cancer, transcriptional repression of BRCA1 may occur more frequently by over-expressed microRNA-182 than by hypermethylation of the BRCA1 promoter.

Transcription in eukaryotic cells requires at the first level of control the choice of RNA polymerases. Eukaryotes have three different RNA polymerases, polymerases I, II and III. In vivo and in vitro experiments have shown that the RNA polymerase II initiates transcription from a site located near a conserved 8 to 10 nucleotide region, 25 to 30 nucleotides upstream (5') from the RNA start site. The nucleotides TATA which are strongly conserved at this site gives the name TATA box or Goldberg-Hogness box. There are other sequences equally important in regulating the access polymerases to the DNA.

Transcription factors play a key role in helping RNA polymerase to bind to promoter region. Transcription factors bind to various consensus sequences like TATA box in Figure. TATA binding protein recognises this box and recruits RNA polymerase II to promoter site for transcription initiation. Transcription factors called activators bind to enhancers. Enhancers lie far from the transcription start site and are position insensitive, while the promoter sites lie close to transcription start site and are position sensitive. Genes lacking TATA box have initiator sequence for initiating transcription. Combination of specific transcription factors is critical for the transcription of a particular gene. Insulators are DNA sequences regulating gene expression. They play a key role in enhancer blocking. Positional enhancer blocking occurs only when insulator lies in between enhancer and operator. Insulators allow genes to remain independent in the genome by restraining the unnecessary signals influencing the gene.

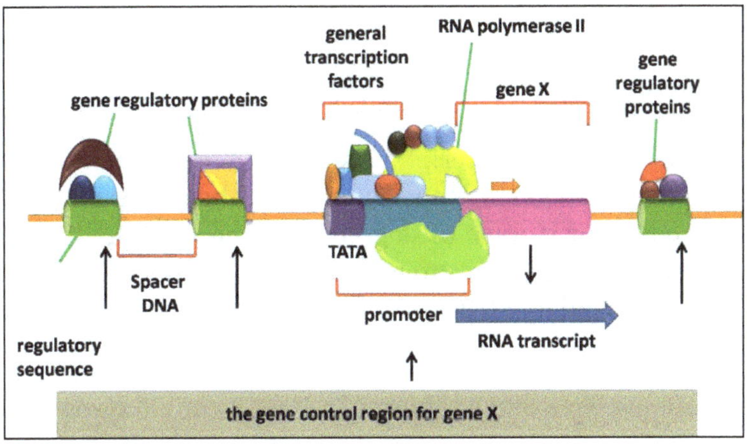

Factors that influence the regulation of a typical gene X.

A variety of structural motifs present in transcription factors interact with DNA at specific sequences. On the basis of DNA binding domains they accommodate, eukaryotic transcription factors are classified as.

Zinc Finger proteins which have regions folding around a central Zn^{2+} ion. Generally DNA binding domains of transcription factors contain such regions. Proteins which do not bind to DNA also show Zn finger patterns.

Leucine Zipper proteins have a leucine at every seventh position on the protein sequence. Such proteins function as transcription factors when present in the form of a dimer. The presence of Leucine helps in parallel zippering of dimer. Yeast Gcn4 is an example of such protein.

Helix turn helix motif is composed of two alpha helices connected by a short amino acid strand. One of them is an amino terminal helix and the other a C- terminal helix. The C-terminal helix is critical for DNA recognition while the other helix stabilizes the DNA- protein interaction. The C-terminal helix binds DNA at major grooves through hydrogen bonding. The Cro, lambda repressor and Catabolite gene activator protein are examples of Helix turn helix motif structures.

Hormones play a crucial role in gene transcription. Type 1 hormones like gluco corticoids being hydrophobic bind to intracellular receptors located in cytoplasm. Hormone receptor interaction causes dissociation of the nuclear localisation signal of receptor which is then transported to nucleus where it binds to hormone response elements to cause transcription of gene. Type II hormones like epinephrine being hydrophilic bind to cell surface receptors and activate G- proteins, leading to gene transcription.

Post-transcriptional Regulation

Post-transcriptional regulation is the control of gene expression at the RNA level, therefore between the transcription and the translation of the gene.

Mechanism

After being produced, the stability and distribution of the different transcripts is regulated (post-transcriptional regulation) by means of RNA binding protein (RBP) that control the various steps and rates cripts: events such as alternative splicing, nuclear degradation (exosome), processing, nuclear export (three alternative pathways), sequestration in P-bodies for storage or degradation and ultimately translation. These proteins achieve these events thanks to a RNA recognition motif (RRM) that binds a specific sequence or secondary structure of the transcripts, typically at the 5' and 3' UTR of the transcript.

Modulating the capping, splicing, addition of a Poly(A) tail, the sequence-specific nuclear export rates and in several contexts sequestration of the RNA transcript occurs in eukaryotes but not in prokaryotes. This modulation is a result of a protein or transcript which in turn is regulated and may have an affinity for certain sequences:

- Capping changes the five prime end of the mRNA to a three prime end by 5'-5' linkage, which protects the mRNA from 5' exonuclease, which degrades foreign RNA. The cap also helps in ribosomal binding.
- Splicing removes the introns, noncoding regions that are transcribed into RNA, in order to make the mRNA able to create proteins. Cells do this by spliceosomes binding on either side of an intron, looping the intron into a circle and then cleaving it off. The two ends of the exons are then joined together.
- Addition of poly(A) tail otherwise known as polyadenylation. That is, a stretch of RNA that is made solely of adenine bases is added to the 3' end, and acts as a buffer to the 3' exonuclease in order to increase the half life of mRNA. In addition, a long poly(A) tail can increase translation. Poly(A)-binding protein (PABP) binds to a long poly(A) tail and mediates the interaction between EIF4E and EIF4G which encourages the initiation of translation.
- RNA editing is a process which results in sequence variation in the RNA molecule, and is catalyzed by enzymes. These enzymes include the Adenosine Deaminase Acting on RNA (ADAR) enzymes, which convert specific adenosine residues to inosine in an mRNA molecule by hydrolytic deamination. Three ADAR enzymes have been cloned, ADAR1, ADAR2 and ADAR3, although only the first two subtypes have been shown to have RNA editing activity. Many mRNAs are vulnerable to the effects of RNA editing, including the glutamate receptor subunits GluR2, GluR3, GluR4, GluR5 and GluR6 (which are components of the AMPA and kainate receptors), the serotonin2C receptor, the GABA-alpha3 receptor subunit, the tryptophan hydroxlase enzyme TPH2, the hepatitis delta virus and more than 16% of microRNAs. In addition to ADAR enzymes, CDAR enzymes exist and these convert cytosines in specific RNA molecules, to uracil. These enzymes are termed 'APOBEC' and have genetic loci at 22q13, a region

close to the chromosomal deletion which occurs in velocardiofacial syndrome (22q11) and which is linked to psychosis. RNA editing is extensively studied in relation to infectious diseases, because the editing process alters viral function.

- mRNA Stability can be manipulated in order to control its half-life, and the poly(A) tail has some effect on this stability, as previously stated. Stable mRNA can have a half life of up to a day or more which allows for the production of more protein product; unstable mRNA is used in regulation that must occur quickly.

microRNA Mediated Regulation

MicroRNAs (miRNAs) appear to regulate the expression of more than 60% of protein coding genes of the human genome. If an miRNA is abundant it can behave as a "switch", turning some genes on or off. However, altered expression of many miRNAs only leads to a modest 1.5- to 4-fold change in protein expression of their target genes. Individual miRNAs often repress several hundred target genes. Repression usually occurs either through translational silencing of the mRNA or through degradation of the mRNA, via complementary binding, mostly to specific sequences in the 3' untranslated region of the target gene's mRNA. The mechanism of translational silencing or degradation of mRNA is implemented through the RNA-induced silencing complex (RISC).

Feedback in the Regulation of RNA Binding Proteins

In metazoans and bacteria, many genes involved in post-post transcriptional regulation are regulated post transcriptionally. For Drosophila RBPs associated with splicing or nonsense mediated decay, analyses of protein-protein and protein-RNA interaction profiles have revealed ubiquitous interactions with RNA and protein products of the same gene. It remains unclear whether these observations are driven by ribosome proximal or ribosome mediated contacts, or if some protein complexes, particularly RNPs, undergo co-translational assembly.

Significance

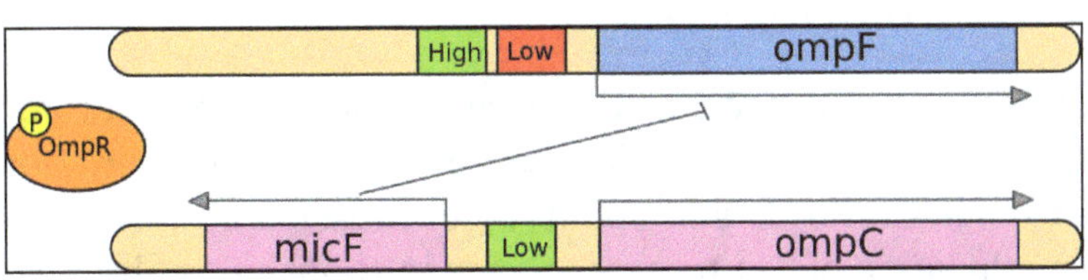

A prokaryotic example: *Salmonella enterica* (a pathogenic γ-proteobacterium) can express two alternative porins depending on the external environment (gut or murky water), this system involves EnvZ (osomotic sensor) which activates OmpR (transcription factor) which can bind to a high affinity promoter even at low concentrations and the low affinity promoter only at high concentrations (by definition): when the concentration of this transcription factor is high it activates OmpC and micF and inhibits OmpF, OmpF is further inhibited post-transcriptionally by micF RNA which binds to the OmpF transcript.

This area of study has recently gained more importance due to the increasing evidence that post-transcriptional regulation plays a larger role than previously expected. Even though protein with DNA binding domains are more abundant than protein with RNA binding domains, a recent study by Cheadle et al. (2005) showed that during T-cell activation 55% of significant changes at the steady-state level had no corresponding changes at the transcriptional level, meaning they were a result of stability regulation alone.

Furthermore, RNA found in the nucleus is more complex than that found in the cytoplasm: more than 95% (bases) of the RNA synthesized by RNA polymerase II never reaches the cytoplasm. The main reason for this is due to the removal of introns which account for 80% of the total bases. Some studies have shown that even after processing the levels of mRNA between the cytoplasm and the nucleus differ greatly.

Developmental biology is a good source of models of regulation, but due to the technical difficulties it was easier to determine the transcription factor cascades than regulation at the RNA level. In fact several key genes such as nanos are known to bind RNA but often their targets are unknown. Although RNA binding proteins may regulate post transcriptionally large amount of the transcriptome, the targeting of a single gene is of interest to the scientific community for medical reasons, this is RNA interference and microRNAs which are both examples of posttranscriptional regulation, which regulate the destruction of RNA and change the chromatin structure. To study post-transcriptional regulation several techniques are used, such as RIP-Chip (RNA immunoprecipitation on chip).

microRNA Role in Cancer

Deficiency of expression of a DNA repair gene occurs in many cancers. Altered microRNA (miRNA) expression that either decreases accurate DNA repair or increases inaccurate microhomology-mediated end joining (MMEJ) DNA repair is often observed in cancers. Deficiency of accurate DNA repair may be a major source of the high frequency of mutations in cancer. Repression of DNA repair genes in cancers by changes in the levels of microRNAs may be a more frequent cause of repression than mutation or epigenetic methylation of DNA repair genes.

For instance, BRCA1 is employed in the accurate homologous recombinational repair (HR) pathway. Deficiency of BRCA1 can cause breast cancer. Down-regulation of BRCA1 due to mutation occurs in about 3% of breast cancers. Down-regulation of BRCA1 due to methylation of its promoter occurs in about 14% of breast cancers. However, increased expression of miR-182 down-regulates BRCA1 mRNA and protein expression, and increased miR-182 is found in 80% of breast cancers.

In another example, a mutated constitutively (persistently) expressed version of the oncogene c-Myc is found in many cancers. Among many functions, c-Myc negatively regulates microRNAs miR-150 and miR-22. These microRNAs normally repress expression of two genes essential for MMEJ, Lig3 and Parp1, thereby inhibiting this inaccurate, mutagenic DNA repair pathway. Muvarak et al. showed, in leukemias, that constitutive expression of c-Myc, leading to down-regulation of miR-150 and miR-22, allowed increased expression of Lig3 and Parp1. This generates genomic instability through increased inaccurate MMEJ DNA repair, and likely contributes to progression to leukemia.

To show the frequent ability of microRNAs to alter DNA repair expression, Hatano et al. performed a large screening study, in which 810 microRNAs were transfected into cells that were then subjected to ionizing radiation (IR). For 324 of these microRNAs, DNA repair was reduced (cells were killed more efficiently by IR) after transfection. For a further 75 microRNAs, DNA repair was increased, with less cell death after IR. This indicates that alterations in microRNAs may often down-regulate DNA repair, a likely important early step in progression to cancer.

Post Transcriptional Control

Post transcriptional control comes into play after the primary transcript is produced. Alternative mRNA splicing is one of the post transcriptional regulatory methods and follows tissue specific or developmental fashion. Tissue specific regulation results in the formation of two different proteins from the same gene in different cells while differentiation specific regulation results in the formation of more than one functionally different protein in the same cell. The best example is the production of calcitonin in thyroid cells, but calcitonin related peptide in brain cells due to alternative splicing. Alternative splicing results in different combinations of exons producing different proteins from the same transcript. Alternative splicing has a switching function which leads to functional or inactive proteins.

Translational Control

This mechanism provides hindrance to protein synthesis through proteins involved in translation. This includes stability of mRNA, probability of translation initiation apart from regulation of overall protein synthesis. These mechanisms control the amount of protein produced from mRNA during a translational event.

Stability of mRNA is important step that controls protein synthesis. Usually the eukaryotic mRNA survives for 3 hours but in certain cases where the amount of protein required in higher amounts, the mRNA survives for several days. In silk worms which produce fibroin, the amount of fibroin for cocoon formation takes days thus the fibroin mRNA survives for several days.Repeated translation of mRNA results in the formation of 10^5 fibroin molecules from 10^4 fibroin mRNAs. Oviduct cells synthesize ovalbumin in chickens but such cells have single copy of this gene per haploid dose. Thus in such circumstances the lifetime of mRNA should be high so that same mRNA is translated repeatedly to achieve the desired protein level.

Reticulocytes synthesize a single type of protein. In such circumstance translation is regulated by modulating total protein synthesis. Globin synthesis in rabbit reticulosites is controlled at the level of translation. Such cells lack nucleus and thus cannot be controlled at the transcriptional level. The process of translation requires elongation factor eIF2 and eIF2 stimulating protein (ESP) besides other molecular players that regulate the event.

Post Translational Control

This control operates after the protein is synthesised in the cell. Post translational modifications include covalent modifications which occur in proteins after translation. Such modifications add another layer of complexity to the complexity embedded in the genome. Post translational modifications play a critical role in regulating activity, localization, and enabling inter molecular interactions inside the cell.

Glysosylation regulates protein stability, influences protein folding kinetics. Ubiquitylation of proteins acts as a signal for proteasomal degradation of proteins. Ubiquitin is a composed of 76 amino acid residues and links to proteins via isopeptide bond. S-nitrosylation is another vital posttranslational modification which modulates protein stability. Caspases residing in the outer mitochondrial membrane are present as S-nitrothiols. On apoptotic signalling, they translocate to the cytosol, become denitrosylated and are activated in the cytosol. Post translational modifications like methylation, acetylation are also involved in diseased states.

We thus understand that, gene expression can be controlled intricately by targeting either the transcription machinery or the translation machinery in a spatio temporal manner. Controls exercised at post transcription and post translation levels have vital implications during disease signalling. This takes us to an interesting logical question on whether apart from the controls exercise at theses states can be totally shut off or activate a gene to initiate favorable responses inside the cell.

Genetic Switches

This is the right time to discuss an interesting experiment designed by Andr'e Lwoff, Francis Jacob and Jacques Monod at the Pasteur institute in Paris nearly 5 decades back. Their experiment showed that a strain of the bacterium E.coli irradiated with ultraviolet light halted their growth and after nearly 90 minutes lyses releasing a crop of viruses into the culture medium. These viruses (originally called lambda) are also called bacteriophages. In many bacteria the lambda virus is dormant but several other bacteria infected by the virus lyse, producing new phage. The normal growth and division when repeated produces a crop of new phages. This experiment clearly demonstrated that the virus switches between two states from the dormant state in the dividing bacterium to the activated state in the bacterium irradiated with UV light. This is a striking example of turning 'ON' or 'OFF' specific genes. We know that genes are the functional components of a living cell. Such living cells (bacterial or human) utilise only a subset of their genes to signal the production of other molecules. Therefore those genes which are expressed are termed 'ON' and those not expressed are turned 'OFF'. In other words we call this phenomenon regulation of gene expression.

This regulation of gene expression as an event takes place not only during the developmental cycle but throughout the life time of a diffrentiative cell.

Gene regulatory proteins and the specific sequences of DNA recognised by these proteins form the basic components of 'genetic switches'. It has been shown that nearly 80% of genctic material of gene switches alters the function of a gene. These genetic switches could be activated or deactivated by external signals, toxins, medium deprivation, stress etc. For example, continuous exposure to sunlight changes the colour of skin cells. Sunlight does not change the structure of the pigmentation gene but alters its function by turning the gene 'ON'. The brilliant discovery of gene switching ON and OFF in lambda phage revolutionised several new experiments in the field and established that the switching mechanism observed in E.coli and lambda phage also applied to eukaryotic cells.

Typical Gene Switching Systems in Cells

Skeletal muscle is composed of two types of fibers- slow twitch fibers which are innately vascular and fast twitch fibres which are deficient in blood vessels. In one of the clinical complications called Critical limb ischemia, blood flow to skeletal muscle is blocked leading to muscle wasting and eventually to the amputation of the limbs. Experiments have established that the genetic switch estrogen related receptor gamma (ERR gamma) when expressed in fast twitch fibers converts them into slow twitch fibers resulting in a significant increase in blood supply to the skeletal muscle. This is a classic case of treating a clinical complication without a pharmacological intervention.

TORC2 is a protein which promotes gluconeogenesis in liver under hypoglycemic conditions by functioning as a metabolic switch. This protein resides outside the nucleus under normal conditions but upon oxidative stress or starvation, shuttles to nucleus and activates a network of genes, vital for handling the insult. Mutations in the TORC protein have been shown to reduce the life expectancy. TORC mutated flies lose their lipid storing capability and have been shown to regain starvation and stress resistance when TORC is expressed in the nervous system.

Experiments on hypoxia tolerant fruit flies have shown that hairy, a transcriptional suppressor is critical for cell survival under hypoxic conditions. Hairy gene may shut off or hinder activation of many genes. On activation it restrains various signalling pathways allowing the cells to circumvent hypoxia. It activates a sort of cutback mechanism in cells culminating in energy conservation which is then used for important functions. Let us now, discuss genetic switches in bacteria.

Bacterial Response to a Single Signal- The Tryptophan Repressor as a Bacterial Switch

The genome of E.coli encodes approximately 4,200 proteins. Its chromosome comprises 4.6×10^6 bp. The expression of many of the genes in E.coli is heavily dependent on the availability of nutrients in the environment.

Tryptophan is a rare amino acid and is a precursor for niacin in eukaryotes. In bacteria, indole is formed from tryptophan. In plants it acts as precursor for biosynthesis the plant hormone auxins. Tryptophan is synthesized from chorismate in five steps catalyzed by three different enzymes which are produced by 5 genes as in Figure. The five genes include TrpE, TrpD, TrpC, TrpB, TrpA. Upstream to TrpE lies TrpL, operator, promoter and far from this stretch lies TrpR. These genes are arranged adjacent to each other on the chromosome as a single operon. The five genes are transcribed as a single mRNA molecule from a single promoter. When tryptophan in the growth medium enters the cell, the cell does not require these enzymes and therefore shuts off the production of these enzymes. The molecular mechanism of the tryptophan switch has been clearly understood and established. The tryptophan repressor is a member of the Helix-Turn-Helix family in which the promoter and the operator are arranged to facilitate classic switching.

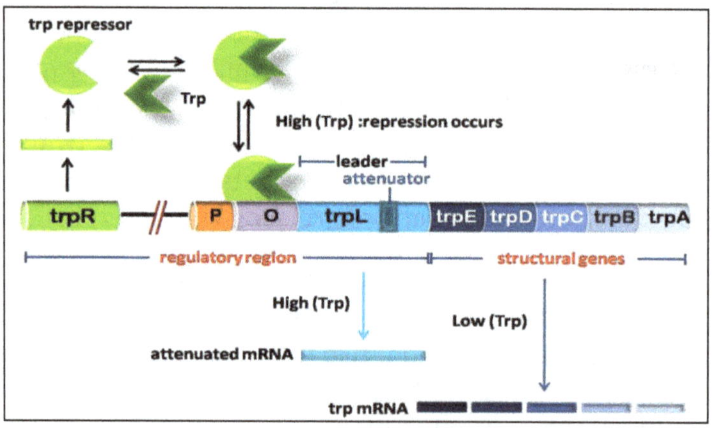

Gene regulation in Trp operon. When the level of trptophan is high: Repression occurs in the circuit. Lower levels of tryptophan results in expression of a gene.

In the first step, chorismate is converted to anthranilate. Anthranilate is converted into phosphoribosylanthranilate which is then converted into carboxy phenyl aminodeoxy ribulose -5-phosphate. After this indole- 3 –glycerol phosphate is formed which is ultimately converted into L-tryptophan via indole formation.

TrpL is a leader sequence and contains a critical region called TrpA meant for attenuation. TrpR produces inactive repressor called apo repressor which can't bind the operator tryptophan acts as corepresssor and activates apo repressor for shutting down genes. The repressor and apo repressor are dimers made up of identical Helix-Turn-Helix monomers. Tryptophan binding prepares the repressor for precise interaction with operator sequences. Trp repressor weakly regulates its own synthesis by binding with an operator site located in its promoter.

In a tryptophan switch, gene expression is regulated through a novel but simple mechanism. In order to bind to the operator DNA, the repressor protein should bind to the amino acid tryptophan through two of its molecules. The binding of tryptophan realigns the Helix-Turn-Helix motif of the repressor presenting it to the major groove of the DNA. When tryptophan is not present, the motif swings inward preventing the binding of the protein to the operator. Thc tryptophan repressor and the operator thus form an elegant switching device that turns gene ON and OFF.

Tryptophan repressor is an example of a negative repressible operon. (Negative with reference to repressor and repressible with reference to tryptophan). If the tryptophan is meager, bacterial structural genes necessary for converting chorismate to tryptophan are activated as the repressor made by TrpR is inactive and is unable to bind to the operator.

In the presence of tryptophan, structural genes need not transcribe as this may result in energy wastage. Tryptophan binds the inactive repressor making it functional.The active repressor binds to operator providing hindrance for RNA polymerase binding. This results in gene repression.

Attenuation

Premature termination of primary transcript in the leader region i.e. before the first structural genes is called attenuation. Attenuation is carried out by attenuator, a sequence within leader region of the tryptophan operon. At this site, choice is made by RNA polymerase either to terminate or continue transcription. Mutants with small deletions in this region produce tryptophan synthesizing enzymes even in the presence of tryptophan.

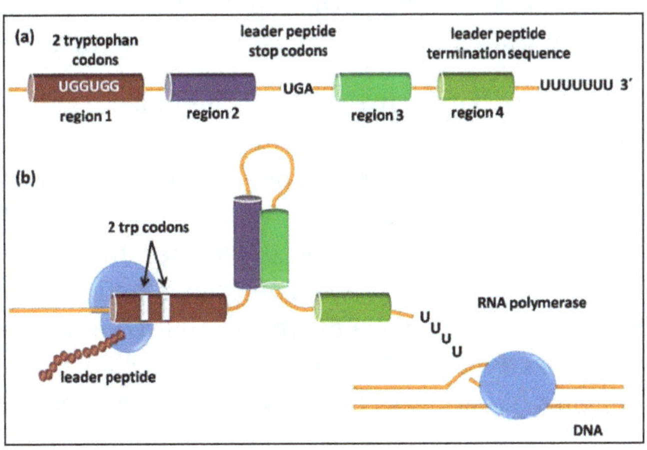

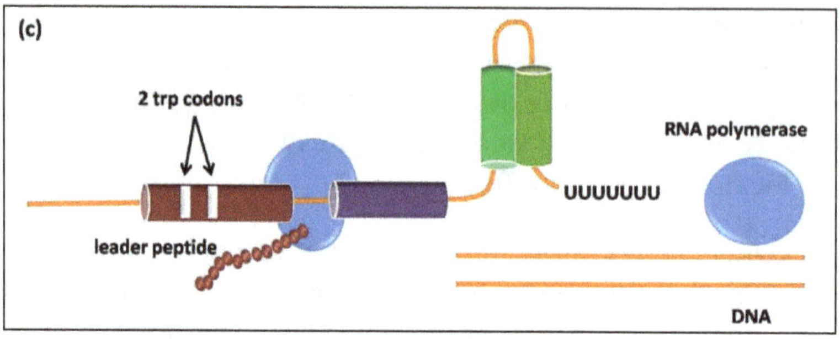

Termination of Transcription regulated by attenuation
(a) Stem-loop structures of the trp operon in the mRNA;
(b) Low level of trp full length mRNA made;
(c) High level transcription of the trp operon is prematurely halted.

The Concept of the lac operon- Transcription Activation and Repression Control

Our earlier discussion on genetic switches gave us an idea of how genetic switches function in the cell and how they elegantly control cellular functions. We discussed the specific case of the simple tryptophan repressor which functions as a switch in bacterial systems. Let us now shift our discussions to complicated types of switching circuits in nature. The complexities in switching in such circuits are brought about by positive and negative controls in the circuit. The classic example of such a switching is the lac operon. The lac operon in E.coli is regulated by the lac repressor and CAP (Catabolite Activator Protein) and is controlled both negatively and positively at the transcriptional state.

Lactose Operon

The lac operon consists of a cluster of functionally related genes controlled by a single promoter. This operon includes the promoter and operator apart from three structural genes lacZ, lacy and lacA. LacZ codes for beta galactosidase, lacY codes for permease and lacA codes for transacetylase. Figure (a) explains the gene regulation in Lac operon. Here Beta galactosidase is a cytoplasmic protein which hydrolyses lactose, Permease acts as lactose importer, and transacetylase detoxifies toxic beta galactosides.

Lactose is a disaccharide which upon hydrolysis forms glucose and galactose in the cell and eventually to break it down while the CAP facilitates bacteria to utilise lactose in the absence of glucose. In the absence of lactose, it is not required for CAP to induce the expression of the lac operon. Here the lac repressor ensures that the lac operon is shut off. This facilitates the controlled region of lac operon to integrate two different signals, ensuring that the operon is expressed only when both conditions are met- lactose present and glucose absent. The other three possible combinations of signals keep the cluster of genes in the OFF state. The lac operon codes for proteins that transport lactose into small amount of permease is found even under repressive condition. The regulator produces a repressor which binds to operator and prevents RNA polymerase from binding to promotor and thus inhibits transcription of three structural genes. The gene regulation process in lac operon is illustrated in Figure (b).

(a)

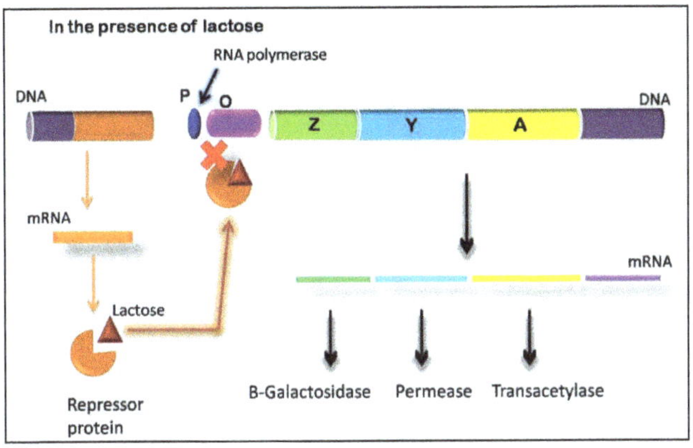

(b)

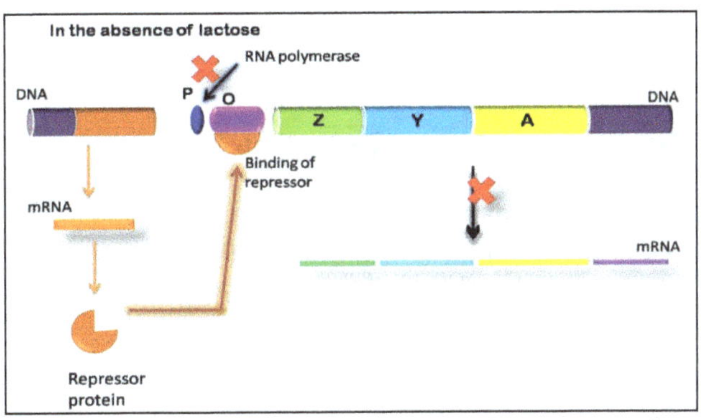

Gene regulation through the Lac operon (a) In the presence of Lactose; (b) In the absence of Lactose.

When both Glucose and Lactose are Absent

Since the lactose (inducer) is not available to bind to the repressor protein, hence the repressor binds to the promoter region and terminates the process of transcription. As a result there is no gene expression.

When both Glucose and Lactose are Present

Under a such condition, the bacteria prefer glucose and utilise lactose only when glucose is exhausted, thus recording two growth curves. This is called diauxic growth.

When Lactose is Present and Glucose is Absent

Lactose is taken in with the help of permease and is converted into allo-lactose. Allolactose binds to the repressor and makes it non functional and thus the three structural genes are transcribed.

Thus the lac– operon is an example of a negative inducible operon -negative with reference to effect of repressor on transcription of structural genes and inducible with reference to effect of lactose on structural gene transcription. Under this situation the levels of cAMP are high.

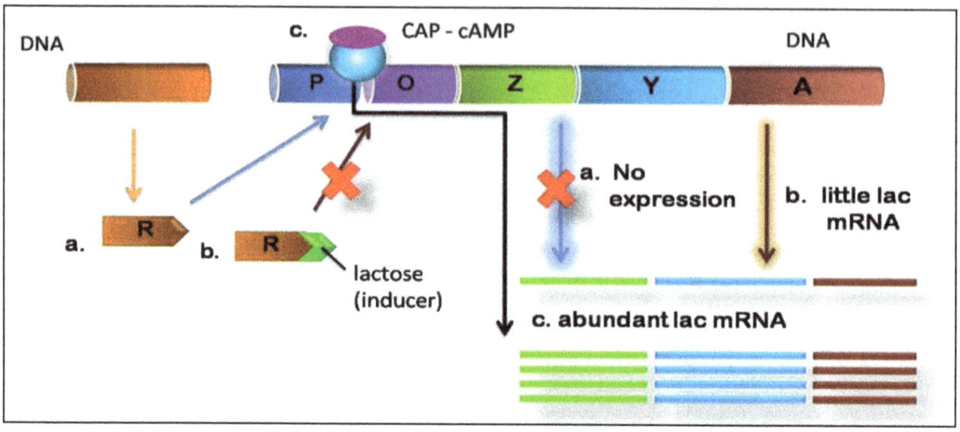

a. Glucose present (cAMP low); no lactose; no lac mRNA.
b. Glucose present (cAMP low); lactose present.
c. No glucose present (cAMP high); lactose present.

Riboswitches- Regulatory Functions of RNA

The central dogma of life has always portrayed the nucleic acids as the blue print for a cell assigning the regulatory and enzymatic functions to the proteins sysnthesised in the cell. Recent work on RNA over the past decade and explorations of its newer functions in the cell strongly challenge the text book view of the central dogma. Though RNA molecules have been shown to be involved in cleavage, splicing and translation and novel gene regulatory mechanisms operating at both the DNA and mRNA level have been explored in detail. Newer functions of specific RNAs that can function as sensors of vitamin B1, B2 and B12 cofactors, have taken center stage. Riboswitches are a fascinating type of RNA structures that regulate gene expression both at the transcription and translation levels by binding to small molecules (ligands). These are structures that form in a messenger RNA and are involved predominantly in gene regulation events in bacteria.

These riboswitches regulate gene expression through the formation of alternative structures which either prematurely terminate transcription or inhibit the initiation of translation when they are in the repressing conformation. Riboswitches regulate the synthesis of Riboflavin, Thiamin and Cobalamin and the metabolism of Methionine, Lysine and Purines. Riboswitches are present in bacterial species, in fungi and in plants. More than 2% of the *Bacillus subtilis* genome has been shown to be regulated by riboswitches.

What do we know about Riboswitches?

Riboswitches fold into compact RNA secondary structures which comprise a base stem, a central multi loop and several branching hair pins as shown in Figure.

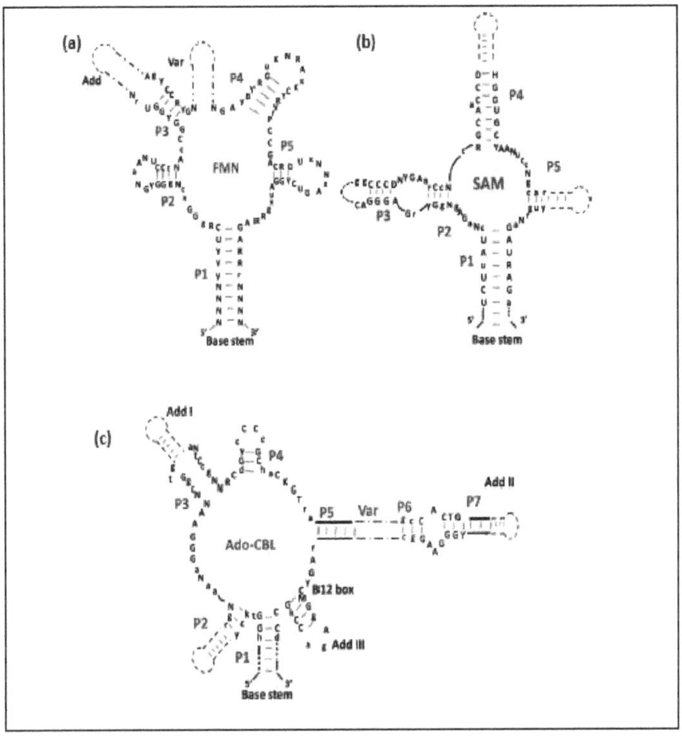

Structure of riboswitches; (a) RFN-element; (b) G-box; (c) B12-element.

The riboswitches distinguish themselves strikingly form other regulatory systems through two features. The first deals with the fact that riboswitches are present across diverse organisms. For example the THI elements are observed in eubacteria, archea and eukaryotes. The S-boxes, G-boxes and L-boxes are observed in gram-positive bacteria from the *Bacillus, Thermotogale* and *Bacteroidetes* species. The next outstanding feature of the riboswitches is that they regulate diverse processes as riboflavin and transport, thiamin synthesis and transport, purine metabolism and synthesis etc as in Table.

Table	Riboswitches and their properties	
Riboswitches	Functional system	Ligand
RFN-element	Riboflavin biosynthesis and transport.	FMN (flavin mononucleotide)
THI-element	Thiamin biosynthesis ; transport of thiamin and related compounds.	TPP (thiamin pyrophosphate)
B12-element	Cobalamin biosynthesis; transport of cobalamin and related compounds; cobalt transport; cobalamin-independent isoenzymes of cobalamin- dependent enzymes.	Coenzyme B12 (adenosylcobalamin)
S-box	Methionine biosynthesis and transport SAM metabolism.	SAM (S- adenosylmethionine)

Mechanism of Function of Riboswitches

As stated earlier, regulation through riboswitches involves the formation of alternative structures.

Condition 1: Repressed State

During repression two RNA structures are formed, the small molecule ligand binds to the structure and stabilizes the switch forming the regulatory hairpin. This hairpin sequesters to the ribosome binding site and can function either to terminate the transcription process or to inhibit the initiation of translation.

Condition 2: Derepressed State

During the derepressed state, the riboswitches is not involved in ligand binding and therefore forms an alternative structure comprising the complementary regions in the riboswitches base stems and a portion of the regulatory hairpin. When the riboswitches directly sequesters the site of translation initiation variations occur in this switch. On change of parity the riboswitches functions as an alternative to the regulatory hairpin, activating gene expression in the presence of the ligand and repressing gene expression when it is not bound.

How Relevant is the Study of Genetic Switches to Systems Biology?

Molecular Biology techniques and precise genetics over the past decades have established the genetic switch as a leading theme of gene regulation. The availability of vast amounts of data from genomics, proteomics and high throughput experimentation, have enabled biology to move from the component centric paradigm to a systems level quest to understand how specific parts of system function together to carry out complex functions. Systems Biology approaches would help illustrate the factors that regulate the efficiency of the switch. Systems Biology can help in modelling the long range interactions between the regulatory proteins in the network and the cooperatively involved in such tight negative auto regulatory networks. The evolution of the switch is an interesting paradigm to be explored using Systems Biology. A question like how robust is the switch against molecular level alterations, how stochastic is the gene regulatory process under such switching would throw interesting discussions on these themes. The construction of genetic toggle switches in vivo would also be an off shoot of such explorations.

References

- Vihervaara, A; Sistonen, L (15 January 2014). "HSF1 at a glance.". Journal of Cell Science. 127 (Pt 2): 261–6. doi:10.1242/jcs.132605. PMID 24421309
- Bruce Alberts; Alexander Johnson; Julian Lewis; Martin Raff; Keith Roberts; Peter Walter (2007). Molecular Biology of the Cell (Fifth ed.). Garland Science. pp. 1268 pages. ISBN 0-8153-4105-9

- Vogelstein B, Papadopoulos N, Velculescu VE, Zhou S, Diaz LA, Kinzler KW (2013). "Cancer genome landscapes". Science. 339 (6127): 1546–58. doi:10.1126/science.1235122. PMC 3749880. PMID 23539594
- Napolitano, G; Lania, L; Majello, B (May 2014). "RNA polymerase II CTD modifications: how many tales from a single tail.". Journal of cellular physiology. 229 (5): 538–44. doi:10.1002/jcp.24483. PMID 24122273
- Bird A (2002). "DNA methylation patterns and epigenetic memory". Genes Dev. 16 (1): 6–21. doi:10.1101/gad.947102. PMID 11782440
- Weaver, Robert J. (2007). "Part V: Post-transcriptional events". Molecular Biology. Boston: McGraw Hill Higher Education. ISBN 0-07-110216-7
- Farazi TA, Spitzer JI, Morozov P, Tuschl T (2011). "miRNAs in human cancer". J. Pathol. 223 (2): 102–15. doi:10.1002/path.2806. PMC 3069496. PMID 21125669

Bioinformatics and Sequence Analysis

Bioinformatics refers to the interdisciplinary field that uses computation and analysis to understand and interpret biological data, which is an essential part of modern biology and medicine. Sequence analysis, sequence assembly, sequence alignment, BLAST, multiple sequence alignment, etc. fall under its domain. This chapter deals with bioinformatics and its related topics in detail.

Bioinformatics

Bioinformatics is an interdisciplinary field that develops methods and software tools for understanding biological data. As an interdisciplinary field of science, bioinformatics combines computer science, statistics, mathematics, and engineering to analyze and interpret biological data. Bioinformatics has been used for *in silico* analyses of biological queries using mathematical and statistical techniques.

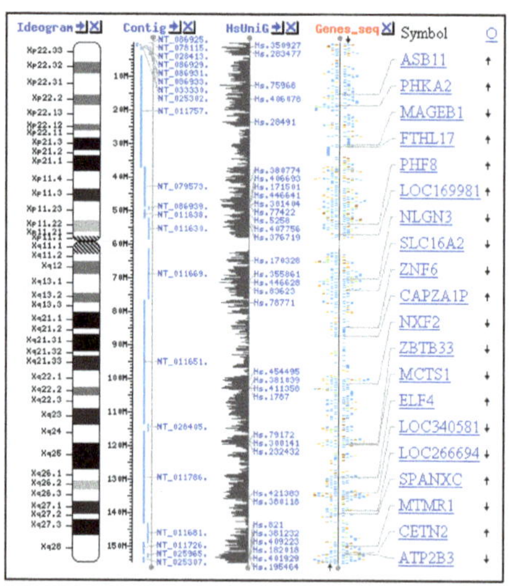

Map of the human X chromosome (from the National Center for Biotechnology Information website).

Bioinformatics is both an umbrella term for the body of biological studies that use computer programming as part of their methodology, as well as a reference to specific analysis "pipelines" that are repeatedly used, particularly in the field of genomics. Common uses of bioinformatics include the identification of candidate genes and nucleotides (SNPs). Often, such identification is made with the aim of better understanding the genetic basis of disease, unique adaptations, desirable properties (esp. in agricultural species), or differences between populations. In a less formal way, bioinformatics also tries to understand the organisational principles within nucleic acid and protein sequences, called proteomics.

Introduction

Bioinformatics has become an important part of many areas of biology. In experimental molecular biology, bioinformatics techniques such as image and signal processing allow extraction of useful results from large amounts of raw data. In the field of genetics and genomics, it aids in sequencing and annotating genomes and their observed mutations. It plays a role in the text mining of biological literature and the development of biological and gene ontologies to organize and query biological data. It also plays a role in the analysis of gene and protein expression and regulation. Bioinformatics tools aid in the comparison of genetic and genomic data and more generally in the understanding of evolutionary aspects of molecular biology. At a more integrative level, it helps analyze and catalogue the biological pathways and networks that are an important part of systems biology. In structural biology, it aids in the simulation and modeling of DNA, RNA, proteins as well as biomolecular interactions.

History

Historically, the term *bioinformatics* did not mean what it means today. Paulien Hogeweg and Ben Hesper coined it in 1970 to refer to the study of information processes in biotic systems. This definition placed bioinformatics as a field parallel to biophysics (the study of physical processes in biological systems) or biochemistry (the study of chemical processes in biological systems).

Sequences

```
5'ATGACGTGGGGA3'
3'TACTGCACCCCT5'
```

Sequences of genetic material are frequently used in bioinformatics and are easier to manage using computers than manually.

Computers became essential in molecular biology when protein sequences became available after Frederick Sanger determined the sequence of insulin in the early 1950s. Comparing multiple sequences manually turned out to be impractical. A pioneer in the field was Margaret Oakley Dayhoff, who has been hailed by David Lipman, director of the National Center for Biotechnology Information, as the "mother and father of bioinformatics." Dayhoff compiled one of the first protein sequence databases, initially published as books and pioneered methods of sequence alignment and molecular evolution. Another early contributor to bioinformatics was Elvin A. Kabat, who pioneered biological sequence analysis in 1970 with his comprehensive volumes of antibody sequences released with Tai Te Wu between 1980 and 1991.

Goals

To study how normal cellular activities are altered in different disease states, the biological data must be combined to form a comprehensive picture of these activities. Therefore, the field of bioinformatics has evolved such that the most pressing task now involves the analysis and interpretation of various types of data. This includes nucleotide and amino acid sequences, protein domains, and protein structures. The actual process of analyzing and interpreting data is referred to as computational biology. Important sub-disciplines within bioinformatics and computational biology include:

- Development and implementation of computer programs that enable efficient access to, use and management of, various types of information.
- Development of new algorithms (mathematical formulas) and statistical measures that assess relationships among members of large data sets. For example, there are methods to locate a gene within a sequence, to predict protein structure and/or function, and to cluster protein sequences into families of related sequences.

The primary goal of bioinformatics is to increase the understanding of biological processes. What sets it apart from other approaches, however, is its focus on developing and applying computationally intensive techniques to achieve this goal. Examples include: pattern recognition, data mining, machine learning algorithms, and visualization. Major research efforts in the field include sequence alignment, gene finding, genome assembly, drug design, drug discovery, protein structure alignment, protein structure prediction, prediction of gene expression and protein–protein interactions, genome-wide association studies, the modeling of evolution and cell division/mitosis.

Bioinformatics now entails the creation and advancement of databases, algorithms, computational and statistical techniques, and theory to solve formal and practical problems arising from the management and analysis of biological data.

Over the past few decades, rapid developments in genomic and other molecular research technologies and developments in information technologies have combined to produce a tremendous amount of information related to molecular biology. Bioinformatics is the name given to these mathematical and computing approaches used to glean understanding of biological processes.

Common activities in bioinformatics include mapping and analyzing DNA and protein sequences, aligning DNA and protein sequences to compare them, and creating and viewing 3-D models of protein structures.

Relation to other Fields

Bioinformatics is a science field that is similar to but distinct from biological computation and computational biology. Biological computation uses bioengineering and biology to build biological computers, whereas bioinformatics uses computation to better understand biology. Bioinformatics and computational biology have similar aims and approaches, but they differ in scale: bioinformatics organizes and analyzes basic biological data, whereas computational biology builds theoretical models of biological systems, just as mathematical biology does with mathematical models.

Analyzing biological data to produce meaningful information involves writing and running software programs that use algorithms from graph theory, artificial intelligence, soft computing, data mining, image processing, and computer simulation. The algorithms in turn depend on theoretical foundations such as discrete mathematics, control theory, system theory, information theory, and statistics.

Sequence Analysis

```
A5ASC3.1  14 SIKLWPPSQTTRLLLVERMANNLST..PSIFTRK..YGSLSKEEARENAKQIEEVACSTANQ.....HYEKEPDGDGGSAVQLYAKECSKLILEVLK 101
B4F917.1  13 SIKLWPPSESTRIMLVDRMTNNLST..ESIFSRK..YRLLGKQEAHENAKTIEELCFALADE.....HFREEPDGDGSSAVQLYAKETSKMMLEVLK 100
A9S1V2.1  23 VFKLWPPSQGTREAVRQKMALKLSS..ACFESQS..FARIELADAQEHARAIEEVAFGAAQE......ADSGGDKTGSAVVMVYAKHASKLMLETLR 109
B9GSN7.1  13 SVKLWPPGQSTRLMLVERMTKNFIT..PSFISRK..YGLLSKEEAEEDAKKIEEVAFAAANQ.....HYEKQPDGDGSSAVQIYAKESSRLMLEVLK 100
Q8H056.1  30 SFSIWPPTQRTRDAVVRRLVDTLGG..DTILCKR..YGAVPAADAEPAARGIEAEAFDAAAA..SGEAAATASVEEGIKALQLYSKEVSRRLLDFVK 120
Q0D4Z3.2  44 SLSIWPPSQRTRDAVVRRLVQTLVA..PSILSQR..YGAVPEAEAGRAAAAVEAEAYAAVTES.SSAAAAPASVEDGIEVLQAYSKEVSRRLLELAK 135
B9MVW8.1  56 SFSIWPPTQRTRDAIISRLIETLST..TSVLSKR..YGTIPKEEASEASRRIEEEAFSGAST.......VASSEKDGLEVLQLYSKEISKRMLETVK 141
Q0IYC5.1  29 SFAVWPPTRRTRDAVVRRLVAVLSGDTTTALRKRYRYGAVPAADAERAARAVEAQAFDAASA....SSSSSSSVEDGIETLQLYSREVSNRLLAFVR 121
A9NW46.1  13 SIKLWPPSESTRLMLVERMTDNLSS..VSFFSRK..YGLLSKEEAAENAKRIEETAFLAAND.....HEAKEPNLDDSSVVQFYAREASKLMLEALK 100
Q9C500.1  57 SLRIWPPTQKTRDAVLNRLIETLST..ESILSKR..YGTLKSDDATTVAKLIEEEAYGVASN.......AVSSDDDGIKILELYSKEISKRMLESVK 142
Q2HRI7.1  25 NYSIWPPKQRTRDAVKNRLIETLST..PSVLTKR..YGTMSADEASAAAIQIEDEAFSVANA.......SSSTSNDNVTILEVYSKEISKRMIETVK 110
Q9M7N3.1  28 SFKIWPPTQRTREAVVRRLVETLTS..QSVLSKR..YGVIPEEDATSAARIIEEEAFSVASV.ASAASTGGRPEDEWIEVLHIYSQEIXQRVVESAK 119
Q9M7N6.1  25 SFSIWPPTQRTRDAVINRLIESLST..PSILSKR..YGTLPQDEASETARLIEEEAFAAAGS.......TASDADDGIEILQVYSKEISKRMIDTVK 110
Q9LE82.1  14 SVKMWPPSKSTRLMLVERMTKNITT..PSIFSRK..YGLLSVEEAEQDAKRIEDLAFATANK.....HFQNEPDGDGTSAVHVYAKESSKLMLDVIK 101
Q9M651.2  13 SIKLWPPSLPTRKALIERITNNFSS..KTIFTEK..YGSLTKDQATENAKRIEDIAFSTANQ.....QFEREPDGDGGSAVQLYAKECSKLILEVLK 100
B9R748.1  48 SLSIWPPTQRTRDAVITRLIETLSS..PSVLSKR..YGTISHDEAESAARRIEDEAFGVANT.......ATSAEDDGLEILQLYSKEISRRMLDTVK 133
```

The sequences of different genes or proteins may be aligned side-by-side to measure their similarity. This alignment compares protein sequences containing WPP domains.

Since the Phage Φ-X174 was sequenced in 1977, the DNA sequences of thousands of organisms have been decoded and stored in databases. This sequence information is analyzed to determine genes that encode proteins, RNA genes, regulatory sequences, structural motifs, and repetitive sequences. A comparison of genes within a species or between different species can show similarities between protein functions, or relations between species (the use of molecular systematics to construct phylogenetic trees). With the growing amount of data, it long ago became impractical to analyze DNA sequences manually. Today, computer programs such as BLAST are used daily to search sequences from more than 260 000 organisms, containing over 190 billion nucleotides. These programs can compensate for mutations (exchanged, deleted or inserted bases) in the DNA sequence, to identify sequences that are related, but not identical. A variant of this sequence alignment is used in the sequencing process itself.

DNA Sequencing

Before sequences can be analyzed they have to be obtained. DNA sequencing is still a non-trivial problem as the raw data may be noisy or afflicted by weak signals. Algorithms have been developed for base calling for the various experimental approaches to DNA sequencing.

Sequence Assembly

Most DNA sequencing techniques produce short fragments of sequence that need to be assembled to obtain complete gene or genome sequences. The so-called shotgun sequencing technique (which was used, for example, by The Institute for Genomic Research (TIGR) to sequence the first bacterial genome, *Haemophilus influenzae*) generates the sequences of many thousands of small DNA fragments (ranging from 35 to 900 nucleotides long, depending on the sequencing technology). The ends of these fragments overlap and, when aligned properly by a genome assembly program, can be used to reconstruct the complete genome. Shotgun sequencing yields sequence data quickly, but the task of assembling the fragments can be quite complicated for larger genomes. For a genome as large as the human genome, it may take many days of CPU time on large-memory, multiprocessor computers to assemble the fragments, and the resulting assembly usually contains numerous gaps that must be filled in later. Shotgun sequencing is the method of choice for virtually all genomes sequenced today, and genome assembly algorithms are a critical area of bioinformatics research.

Genome Annotation

In the context of genomics, annotation is the process of marking the genes and other biological features in a DNA sequence. This process needs to be automated because most genomes are too large to annotate by hand, not to mention the desire to annotate as many genomes as possible, as the rate of sequencing has ceased to pose a bottleneck. Annotation is made possible by the fact that genes have recognisable start and stop regions, although the exact sequence found in these regions can vary between genes.

The first genome annotation software system was designed in 1995 by Owen White, who was part of the team at The Institute for Genomic Research that sequenced and analyzed the first genome of a free-living organism to be decoded, the bacterium *Haemophilus influenzae*. White built a software system to find the genes (fragments of genomic sequence that encode proteins), the transfer RNAs, and to make initial assignments of function to those genes. Most current genome annotation systems work similarly, but the programs available for analysis of genomic DNA, such as the GeneMark program trained and used to find protein-coding genes in *Haemophilus influenzae*, are constantly changing and improving.

Following the goals that the Human Genome Project left to achieve after its closure in 2003, a new project developed by the National Human Genome Research Institute in the U.S appeared. The so-called ENCODE project is a collaborative data collection of the functional elements of the human genome that uses next-generation DNA-sequencing technologies and genomic tiling arrays, technologies able to generate automatically large amounts of data with lower research costs but with the same quality and viability.

Computational Evolutionary Biology

Evolutionary biology is the study of the origin and descent of species, as well as their change over time. Informatics has assisted evolutionary biologists by enabling researchers to:

- Trace the evolution of a large number of organisms by measuring changes in their DNA, rather than through physical taxonomy or physiological observations alone,
- More recently, compare entire genomes, which permits the study of more complex evolutionary events, such as gene duplication, horizontal gene transfer, and the prediction of factors important in bacterial speciation,
- Build complex computational population genetics models to predict the outcome of the system over time,
- Track and share information on an increasingly large number of species and organisms.

Future work endeavours to reconstruct the now more complex tree of life.

The area of research within computer science that uses genetic algorithms is sometimes confused with computational evolutionary biology, but the two areas are not necessarily related.

Comparative Genomics

The core of comparative genome analysis is the establishment of the correspondence between genes (orthology analysis) or other genomic features in different organisms. It is these intergenomic maps that make it possible to trace the evolutionary processes responsible for the divergence of two genomes. A multitude of evolutionary events acting at various organizational levels shape genome evolution. At the lowest level, point mutations affect individual nucleotides. At a higher level, large chromosomal segments undergo duplication, lateral transfer, inversion, transposition, deletion and insertion. Ultimately, whole genomes are involved in processes of hybridization, polyploidization and endosymbiosis, often leading to rapid speciation. The complexity of genome evolution poses many exciting challenges to developers of mathematical models and algorithms, who have recourse to a spectra of algorithmic, statistical and mathematical techniques, ranging from exact, heuristics, fixed parameter and approximation algorithms for problems based on parsimony models to Markov Chain Monte Carlo algorithms for Bayesian analysis of problems based on probabilistic models.

Many of these studies are based on the homology detection and protein families computation.

Pan Genomics

Pan genomics is a concept introduced in 2005 by Tettelin and Medini which eventually took root in bioinformatics. Pan genome is the complete gene repertoire of a particular taxonomic group: although initially applied to closely related strains of a species, it can be applied to a larger context like genus, phylum etc. It is divided in two parts- The Core genome: Set of genes common to all the genomes under study (These are often house-keeping genes vital for survival) and The Dispensable/Flexible Genome: Set of genes not present in all but one or some genomes under study. a bioinformatics tool BPGA can be used to characterize the Pan Genome of bacterial species.

Genetics of Disease

With the advent of next-generation sequencing we are obtaining enough sequence data to map the genes of complex diseases such as diabetes, infertility, breast cancer or Alzheimer's Disease. Genome-wide association studies are a useful approach to pinpoint the mutations responsible for such complex diseases. Through these studies, thousands of DNA variants have been identified that are associated with similar diseases and traits. Furthermore, the possibility for genes to be used at prognosis, diagnosis or treatment is one of the most essential applications. Many studies are discussing both the promising ways to choose the genes to be used and the problems and pitfalls of using genes to predict disease presence or prognosis.

Analysis of Mutations in Cancer

In cancer, the genomes of affected cells are rearranged in complex or even unpredictable ways. Massive sequencing efforts are used to identify previously unknown point mutations in a variety of genes in cancer. Bioinformaticians continue to produce specialized automated systems to manage the sheer volume of sequence data produced, and they create new algorithms and software to compare the sequencing results to the growing collection of human genome sequences and germline polymorphisms. New physical detection technologies are employed, such as oligonucleotide microarrays to identify chromosomal gains and losses (called comparative genomic hybridization), and single-nucleotide polymorphism arrays to detect known *point mutations*. These detection methods simultaneously measure several hundred thousand sites throughout the genome, and when used in high-throughput to measure thousands of samples, generate terabytes of data per experiment. Again the massive amounts and new types of data generate new opportunities for bioinformaticians. The data is often found to contain considerable variability, or noise, and thus Hidden Markov model and change-point analysis methods are being developed to infer real copy number changes.

With the breakthroughs that this next-generation sequencing technology is providing to the field of Bioinformatics, cancer genomics could drastically change. These new methods and software allow bioinformaticians to sequence many cancer genomes quickly and affordably. This could create a more flexible process for classifying types of cancer by analysis of cancer driven mutations in the genome. Furthermore, tracking of patients while the disease progresses may be possible in the future with the sequence of cancer samples.

Another type of data that requires novel informatics development is the analysis of lesions found to be recurrent among many tumors.

Gene and Protein Expression

Analysis of Gene Expression

The expression of many genes can be determined by measuring mRNA levels with multiple techniques including microarrays, expressed cDNA sequence tag (EST) sequencing, serial analysis of gene expression (SAGE) tag sequencing, massively parallel signature sequencing (MPSS), RNA-Seq, also known as "Whole Transcriptome Shotgun Sequencing" (WTSS), or various applications of multiplexed in-situ hybridization. All of these techniques are extremely noise-prone and/or subject to bias in the biological measurement, and a major research area in computational biology involves developing statistical tools to separate signal from noise in high-throughput gene expression studies. Such studies are often used to determine the genes implicated in a disorder: one might compare microarray data from cancerous epithelial cells to data from non-cancerous cells to determine the transcripts that are up-regulated and down-regulated in a particular population of cancer cells.

Analysis of Protein Expression

Protein microarrays and high throughput (HT) mass spectrometry (MS) can provide a snapshot of the proteins present in a biological sample. Bioinformatics is very much involved in making sense of protein microarray and HT MS data; the former approach faces similar problems as with microarrays targeted at mRNA, the latter involves the problem of matching large amounts of mass data against predicted masses from protein sequence databases, and the complicated statistical analysis of samples where multiple, but incomplete peptides from each protein are detected.

Analysis of Regulation

Regulation is the complex orchestration of events by which a signal, potentially an extracellular signal such as a hormone, eventually leads to an increase or decrease in the activity of one or more proteins. Bioinformatics techniques have been applied to explore various steps in this process.

For example, gene expression can be regulated by nearby elements in the genome. Promoter analysis involves the identification and study of sequence motifs in the DNA surrounding the coding region of a gene. These motifs influence the extent to which that region is transcribed into mRNA. Enhancer elements far away from the promoter can also regulate gene expression, through three-dimensional looping interactions. These interactions can be determined by bioinformatic analysis of chromosome conformation capture experiments.

Expression data can be used to infer gene regulation: one might compare microarray data from a wide variety of states of an organism to form hypotheses about the genes involved in each state. In a single-cell organism, one might compare stages of the cell cycle, along with various stress conditions (heat shock, starvation, etc.). One can then apply clustering algorithms to that expression data to determine which genes are co-expressed. For example, the upstream regions (promoters) of co-expressed genes can be searched for over-represented regulatory elements. Examples of clustering algorithms applied in gene clustering are k-means clustering, self-organizing maps (SOMs), hierarchical clustering, and consensus clustering methods.

Analysis of Cellular Organization

Several approaches have been developed to analyze the location of organelles, genes, proteins, and other components within cells. This is relevant as the location of these components affects the events within a cell and thus helps us to predict the behavior of biological systems. A gene ontology category, *cellular compartment*, has been devised to capture subcellular localization in many biological databases.

Microscopy and Image Analysis

Microscopic pictures allow us to locate both organelles as well as molecules. It may also help us to distinguish between normal and abnormal cells, e.g. in cancer.

Protein Localization

The localization of proteins helps us to evaluate the role of a protein. For instance, if a protein is found in the nucleus it may be involved in gene regulation or splicing. By contrast, if a protein is found in mitochondria, it may be involved in respiration or other metabolic processes. Protein localization is thus an important component of protein function prediction.

Chromosome Topology

Data from high-throughput chromosome conformation capture experiments, such as Hi-C (experiment) and ChIA-PET, can provide information on the spatial proximity of DNA loci. Anaylsis of these experiments can determine the three-dimensional structure and nuclear organization of chromatin. Bioinformatic challenges in this field include partitioning the genome into domains, such as Topologically Associating Domains (TADs), that are organised together in three-dimensional space.

Structural Bioinformatics

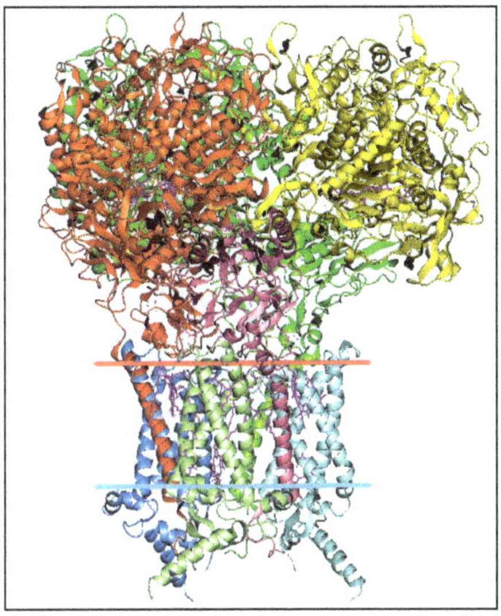

3-dimensional protein structures such as this one are common subjects in bioinformatic analyses.

Protein structure prediction is another important application of bioinformatics. The amino acid sequence of a protein, the so-called primary structure, can be easily determined from the sequence on the gene that codes for it. In the vast majority of cases, this primary structure uniquely determines a structure in its native environment. (Of course, there are exceptions, such as the bovine spongiform encephalopathy – a.k.a. Mad Cow Disease – prion.) Knowledge of this structure is vital in understanding the function of the protein. Structural information is usually classified as one of *secondary*, *tertiary* and *quaternary* structure. A viable general solution to such predictions remains an open problem. Most efforts have so far been directed towards heuristics that work most of the time.

One of the key ideas in bioinformatics is the notion of homology. In the genomic branch of bioinformatics, homology is used to predict the function of a gene: if the sequence of gene *A*, whose function is known, is homologous to the sequence of gene *B,* whose function is unknown, one could infer that B may share A's function. In the structural branch of bioinformatics, homology is used to determine which parts of a protein are important in structure formation and interaction with other proteins. In a technique called homology modeling, this information is used to predict the structure of a protein once the structure of a homologous protein is known. This currently remains the only way to predict protein structures reliably.

One example of this is the similar protein homology between hemoglobin in humans and the hemoglobin in legumes (leghemoglobin). Both serve the same purpose of transporting oxygen in the organism. Though both of these proteins have completely different amino acid sequences, their protein structures are virtually identical, which reflects their near identical purposes.

Other techniques for predicting protein structure include protein threading and *de novo* (from scratch) physics-based modeling.

Network and Systems Biology

Network analysis seeks to understand the relationships within biological networks such as metabolic or protein–protein interaction networks. Although biological networks can be constructed from a single type of molecule or entity (such as genes), network biology often attempts to integrate many different data types, such as proteins, small molecules, gene expression data, and others, which are all connected physically, functionally, or both.

Systems biology involves the use of computer simulations of cellular subsystems (such as the networks of metabolites and enzymes that comprise metabolism, signal transduction pathways and gene regulatory networks) to both analyze and visualize the complex connections of these cellular processes. Artificial life or virtual evolution attempts to understand evolutionary processes via the computer simulation of simple (artificial) life forms.

Molecular Interaction Networks

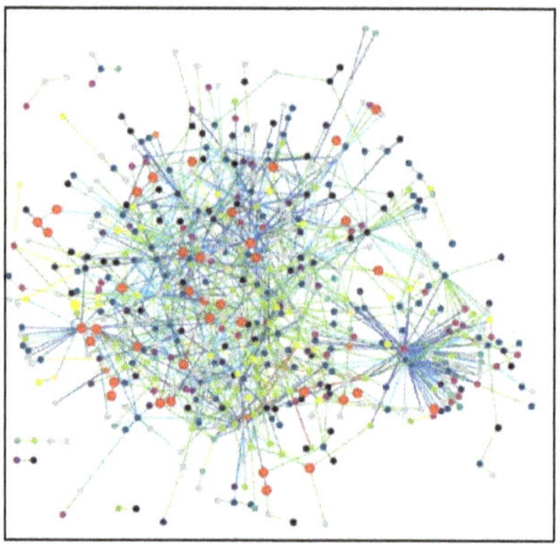

Interactions between proteins are frequently visualized and analyzed using networks. This network is made up of protein–protein interactions from *Treponema pallidum*, the causative agent of syphilis and other diseases.

Tens of thousands of three-dimensional protein structures have been determined by X-ray crystallography and protein nuclear magnetic resonance spectroscopy (protein NMR) and a central question in structural bioinformatics is whether it is practical to predict possible protein–protein interactions only based on these 3D shapes, without performing protein–protein interaction experiments. A variety of methods have been developed to tackle the protein–protein docking problem, though it seems that there is still much work to be done in this field.

Other interactions encountered in the field include Protein–ligand (including drug) and protein–peptide. Molecular dynamic simulation of movement of atoms about rotatable bonds is the fundamental principle behind computational algorithms, termed docking algorithms, for studying molecular interactions.

Others

Literature Analysis

The growth in the number of published literature makes it virtually impossible to read every paper, resulting in disjointed sub-fields of research. Literature analysis aims to employ computational and statistical linguistics to mine this growing library of text resources. For example:

- Abbreviation recognition – identify the long-form and abbreviation of biological terms.
- Named entity recognition – recognizing biological terms such as gene names.
- Protein–protein interaction – identify which proteins interact with which proteins from text.

The area of research draws from statistics and computational linguistics.

High-throughput Image Analysis

Computational technologies are used to accelerate or fully automate the processing, quantification and analysis of large amounts of high-information-content biomedical imagery. Modern image analysis systems augment an observer's ability to make measurements from a large or complex set of images, by improving accuracy, objectivity, or speed. A fully developed analysis system may completely replace the observer. Although these systems are not unique to biomedical imagery, biomedical imaging is becoming more important for both diagnostics and research. Some examples are:

- High-throughput and high-fidelity quantification and sub-cellular localization (high-content screening, cytohistopathology, Bioimage informatics).
- Morphometrics.
- Clinical image analysis and visualization.
- Determining the real-time air-flow patterns in breathing lungs of living animals.
- Quantifying occlusion size in real-time imagery from the development of and recovery during arterial injury.
- Making behavioral observations from extended video recordings of laboratory animals.
- Infrared measurements for metabolic activity determination.
- Inferring clone overlaps in DNA mapping, e.g. the Sulston score.

High-throughput Single Cell Data Analysis

Computational techniques are used to analyse high-throughput, low-measurement single cell data, such as that obtained from flow cytometry. These methods typically involve finding populations of cells that are relevant to a particular disease state or experimental condition.

Biodiversity Informatics

Biodiversity informatics deals with the collection and analysis of biodiversity data, such as taxonomic databases, or microbiome data. Examples of such analyses include phylogenetics, niche modelling, species richness mapping, or species identification tools.

Databases

Databases are essential for bioinformatics research and applications. Many databases exist, covering various information types: for example, DNA and protein sequences, molecular structures, phenotypes and biodiversity. Databases may contain empirical data (obtained directly from experiments), predicted data (obtained from analysis), or, most commonly, both. They may be specific to a particular organism, pathway or molecule of interest. Alternatively, they can incorporate data compiled from multiple other databases. These databases vary in their format, way of accession and whether they are public or not.

Some of the most commonly used databases are listed below. For a more comprehensive list, please check the link at the beginning of the subsection.

- Used in Motif Finding: GenomeNet MOTIF Search.
- Used in Gene Ontology: ToppGene FuncAssociate, Enrichr, GATHER.
- Used in finding Protein Structures/Family: Pfam.
- Used for Next Generation Sequencing: Sequence Read Archive.
- Used in Gene Expression Analysis: GEO, ArrayExpress.
- Used in Network Analysis: Interaction Analysis Databases(BioGRID, MINT, HPRD, Curated Human Signaling Network), Functional Networks (STRING, KEGG).
- Used in design of synthetic genetic circuits: GenoCAD.

Software and Tools

Software tools for bioinformatics range from simple command-line tools, to more complex graphical programs and standalone web-services available from various bioinformatics companies or public institutions.

Open-source Bioinformatics Software

Many free and open-source software tools have existed and continued to grow since the 1980s. The combination of a continued need for new algorithms for the analysis of emerging types of biological readouts, the potential for innovative *in silico* experiments, and freely available open code bases have helped to create opportunities for all research groups to contribute to both bioinformatics and the range of open-source software available, regardless of their funding arrangements. The open source tools often act as incubators of ideas, or community-supported plug-ins in commercial applications. They may also provide *de facto* standards and shared object models for assisting with the challenge of bioinformation integration.

The range of open-source software packages includes titles such as Bioconductor, BioPerl, Biopython, BioJava, BioJS, BioRuby, Bioclipse, EMBOSS, .NET Bio, Orange with its bioinformatics add-on, Apache Taverna, UGENE and GenoCAD. To maintain this tradition and create further opportunities, the non-profit Open Bioinformatics Foundation have supported the annual Bioinformatics Open Source Conference (BOSC) since 2000.

An alternative method to build public bioinformatics databases is to use the MediaWiki engine with the *WikiOpener* extension. This system allows the database to be accessed and updated by all experts in the field.

Web Services in Bioinformatics

SOAP- and REST-based interfaces have been developed for a wide variety of bioinformatics applications allowing an application running on one computer in one part of the world to use algorithms, data and computing resources on servers in other parts of the world. The main advantages derive from the fact that end users do not have to deal with software and database maintenance overheads.

Basic bioinformatics services are classified by the EBI into three categories: SSS (Sequence Search Services), MSA (Multiple Sequence Alignment), and BSA (Biological Sequence Analysis). The availability of these service-oriented bioinformatics resources demonstrate the applicability of web-based bioinformatics solutions, and range from a collection of standalone tools with a common data format under a single, standalone or web-based interface, to integrative, distributed and extensible bioinformatics workflow management systems.

Bioinformatics Workflow Management Systems

A Bioinformatics workflow management system is a specialized form of a workflow management system designed specifically to compose and execute a series of computational or data manipulation steps, or a workflow, in a Bioinformatics application.

Such systems are designed to:

- Provide an easy-to-use environment for individual application scientists themselves to create their own workflows.
- Provide interactive tools for the scientists enabling them to execute their workflows and view their results in real-time.
- Simplify the process of sharing and reusing workflows between the scientists.
- Enable scientists to track the provenance of the workflow execution results and the workflow creation steps.

Some of the platforms giving this service: Galaxy, Kepler, Taverna, UGENE, Anduril.

Education Platforms

Software platforms designed to teach bioinformatics concepts and methods include Rosalind and online courses offered through the Swiss Institute of Bioinformatics Training Portal. The Canadian Bioinformatics Workshops provides videos and slides from training workshops on their website under a Creative Commons license. The 4273π project or 4273pi project also offers open source educational materials for free. The course runs on low cost raspberry pi computers and has been used to teach adults and school pupils. 4273π is actively developed by a consortium of academics and research staff who have run research level bioinformatics using raspberry pi computers and the 4273π operating system.

MOOC platforms also provide online certifications in bioinformatics and related disciplines, including Coursera's Bioinformatics Specialization (UC San Diego) and Genomic Data Science Specialization (Johns Hopkins) as well as EdX's Data Analysis for Life Sciences XSeries (Harvard).

Conferences

There are several large conferences that are concerned with bioinformatics. Some of the most notable examples are Intelligent Systems for Molecular Biology (ISMB), European Conference on Computational Biology (ECCB), and Research in Computational Molecular Biology (RECOMB).

Sequence Analysis

In bioinformatics, sequence analysis is the process of subjecting a DNA, RNA or peptide sequence to any of a wide range of analytical methods to understand its features, function, structure, or evolution. Methodologies used include sequence alignment, searches against biological databases, and others. Since the development of methods of high-throughput production of gene and protein sequences, the rate of addition of new sequences to the databases increased exponentially. Such a collection of sequences does not, by itself, increase the scientist's understanding of the biology of organisms. However, comparing these new sequences to those with known functions is a key way of understanding the biology of an organism from which the new sequence comes. Thus, sequence analysis can be used to assign function to genes and proteins by the study of the similarities between the compared sequences. Nowadays, there are many tools and techniques that provide the sequence comparisons (sequence alignment) and analyze the alignment product to understand its biology.

Sequence analysis in molecular biology includes a very wide range of relevant topics:

1. The comparison of sequences in order to find similarity, often to infer if they are related (homologous).
2. Identification of intrinsic features of the sequence such as active sites, post translational modification sites, gene-structures, reading frames, distributions of introns and exons and regulatory elements.
3. Identification of sequence differences and variations such as point mutations and single nucleotide polymorphism (SNP) in order to get the genetic marker.
4. Revealing the evolution and genetic diversity of sequences and organisms.
5. Identification of molecular structure from sequence alone.

In chemistry, sequence analysis comprises techniques used to determine the sequence of a polymer formed of several monomers. In molecular biology and genetics, the same process is called simply "sequencing".

In marketing, sequence analysis is often used in analytical customer relationship management applications, such as NPTB models (Next Product to Buy).

In sociology, sequence methods are increasingly used to study life-course and career trajectories, patterns of organizational and national development, conversation and interaction structure, and the problem of work/family synchrony. This body of research has given rise to the emerging subfield of social sequence analysis.

History

Since the very first sequences of the insulin protein was characterised by Fred Sanger in 1951 biologists have been trying to use this knowledge to understand the function of molecules. According to Michael Levitt, sequence analysis was born in the period from 1969-1977. In 1969 the analysis of sequences of transfer RNAs were used to infer residue interactions from correlated changes in the nucleotide sequences giving rise to a model of the tRNA secondary structure. In 1970, Saul B. Needleman and Christian D. Wunsch published the first computer algorithm for aligning two sequences. Over this time developments in obtaining nucleotide sequence greatly improved leading to the publication of the first complete genome of a bacteriophage in 1977.

Sequence Alignment

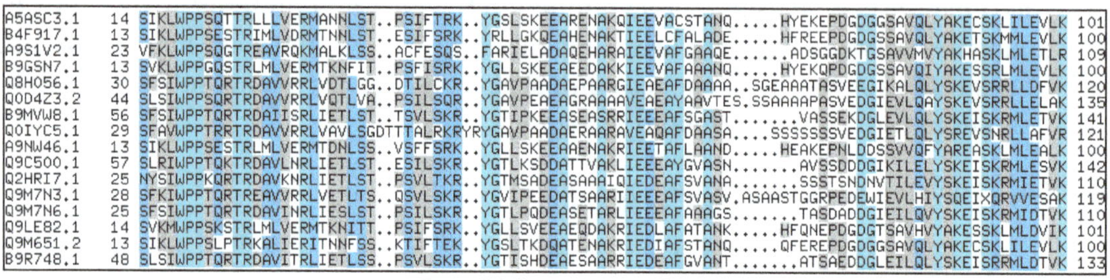

Example multiple sequence alignment.

There are millions of protein and nucleotide sequences known. These sequences fall into many groups of related sequences known as protein families or gene families. Relationships between these sequences are usually discovered by aligning them together and assigning this alignment a score. There are two main types of sequence alignment. Pair-wise sequence alignment only compares two sequences at a time and multiple sequence alignment compares many sequences in one go. Two important algorithms for aligning pairs of sequences are the Needleman-Wunsch algorithm and the Smith-Waterman algorithm. Popular tools for sequence alignment include:

- Pair-wise alignment - BLAST.
- Multiple alignment - ClustalW, PROBCONS, MUSCLE, MAFFT, and T-Coffee.

A common use for pairwise sequence alignment is to take a sequence of interest and compare it to all known sequences in a database to identify homologous sequences. In general the matches in the database are ordered to show the most closely related sequences first followed by sequences with diminishing similarity. These matches are usually reported with a measure of statistical significance such as an Expectation value.

Profile Comparison

In 1987, Michael Gribskov, Andrew McLachlan, and David Eisenberg introduced the method of profile comparison for identifying distant similarities between proteins. Rather than using a single sequence, profile methods use a multiple sequence alignment to encode a profile which contains information about the conservation level of each residue. These profiles can then be used to search collections of sequences to find sequences that are related. Profiles are also known as Position Specific Scoring Matrices (PSSMs). In 1993, a probabilistic interpretation of profiles was introduced by David Haussler and colleagues using hidden Markov models. These models have become known as profile-HMMs.

In recent years, methods have been developed that allow the comparison of profiles directly to each other. These are known as profile-profile comparison methods.

Sequence Assembly

Sequence assembly refers to the reconstruction of a DNA sequence by aligning and merging small DNA fragments. It is an integral part of modern DNA sequencing. Since presently-available DNA sequencing technologies are ill-suited for reading long sequences, large pieces of DNA (such as genomes) are often sequenced by (1) cutting the DNA into small pieces, (2) reading the small fragments, and (3) reconstituting the original DNA by merging the information on various fragment.

Gene Prediction

Gene prediction or gene finding refers to the process of identifying the regions of genomic DNA that encode genes. This includes protein-coding genes as well as RNA genes, but may also include prediction of other functional elements such as regulatory regions. Gene finding is one of the first and most important steps in understanding the genome of a species once it has been sequenced. In general the prediction of bacterial genes is significantly simpler and more accurate than the prediction of genes in eukaryotic species that usually have complex intron/exon patterns.

Protein Structure Prediction

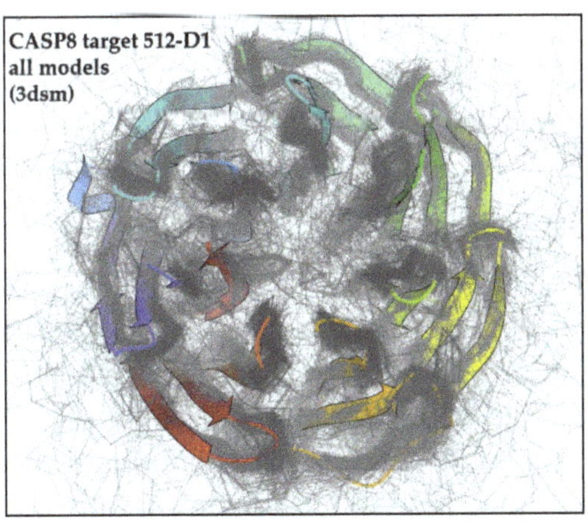

Target protein structure (3dsm, shown in ribbons), with Calpha backbones (in gray) of 354 predicted models for it submitted in the CASP8 structure-prediction experiment.

The 3D structures of molecules are of great importance to their functions in nature. Since structural prediction of large molecules at an atomic level is largely intractable problem, some biologists introduced ways to predict 3D structure at a primary sequence level. This includes biochemical or statistical analysis of amino acid residues in local regions and structural inference from homologs (or other potentially related proteins) with known 3D structures.

There have been a large number of diverse approaches to solve the structure prediction problem. In order to determine which methods were most effective a structure prediction competition was founded called CASP (Critical Assessment of Structure Prediction).

Methodology

The tasks that lie in the space of sequence analysis are often non-trivial to resolve and require the use of relatively complex approaches. Of the many types of methods used in practice, the most popular include:

- DNA patterns.
- Dynamic programming.
- Artificial Neural Network.
- Hidden Markov Model.
- Support Vector Machine.
- Clustering.
- Bayesian Network.
- Regression Analysis.
- Sequence mining.
- Alignment-free sequence analysis.

Sequence Assembly

In bioinformatics, sequence assembly refers to aligning and merging fragments from a longer DNA sequence in order to reconstruct the original sequence. This is needed as DNA sequencing technology cannot read whole genomes in one go, but rather reads small pieces of between 20 and 30000 bases, depending on the technology used. Typically the short fragments, called reads, result from shotgun sequencing genomic DNA, or gene transcript (ESTs).

The problem of sequence assembly can be compared to taking many copies of a book, passing each of them through a shredder with a different cutter, and piecing the text of the book back together just by looking at the shredded pieces. Besides the obvious difficulty of this task, there are some extra practical issues: the original may have many repeated paragraphs, and some shreds may be modified during shredding to have typos. Excerpts from another book may also be added in, and some shreds may be completely unrecognizable.

Genome Assemblers

The first sequence assemblers began to appear in the late 1980s and early 1990s as variants of simpler sequence alignment programs to piece together vast quantities of fragments generated by automated sequencing instruments called DNA sequencers. As the sequenced organisms grew in size and complexity (from small viruses over plasmids to bacteria and finally eukaryotes), the assembly programs used in these genome projects needed increasingly sophisticated strategies to handle:

- Terabytes of sequencing data which need processing on computing clusters;
- Identical and nearly identical sequences (known as *repeats*) which can, in the worst case, increase the time and space complexity of algorithms exponentially;
- Errors in the fragments from the sequencing instruments, which can confound assembly.

Faced with the challenge of assembling the first larger eukaryotic genomes—the fruit fly Drosophila melanogaster in 2000 and the human genome just a year later,—scientists developed assemblers like Celera Assembler and Arachne able to handle genomes of 100-300 million base pairs. Subsequent to these efforts, several other groups, mostly at the major genome sequencing centers, built large-scale assemblers, and an open source effort known as AMOS was launched to bring together all the innovations in genome assembly technology under the open source framework.

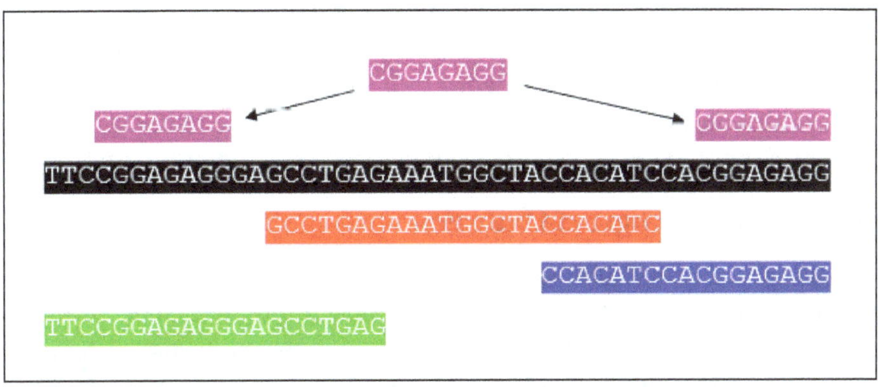

Sample sequence showing how a sequence assembler would take fragments and match by overlaps. Image also shows the potential problem of repeats in the sequence.

EST Assemblers

Expressed Sequence Tag or EST assembly differs from genome assembly in several ways. The sequences for EST assembly are the transcribed mRNA of a cell and represent only a subset of the whole genome. At a first glance, underlying algorithmical problems differ between genome and EST assembly. For instance, genomes often have large amounts of repetitive sequences, mainly in the inter-genic parts. Since ESTs represent gene transcripts, they will not contain these repeats. On the other hand, cells tend to have a certain number of genes that are constantly expressed in very high numbers (housekeeping genes), which again leads to the problem of similar sequences present in high numbers in the data set to be assembled.

Furthermore, genes sometimes overlap in the genome (sense-antisense transcription), and should ideally still be assembled separately. EST assembly is also complicated by features like (cis-) alternative splicing, trans-splicing, single-nucleotide polymorphism, and post-transcriptional modification.

De-novo vs. Mapping Assembly

In sequence assembly, two different types can be distinguished:

1. De-novo: assembling short reads to create full-length (sometimes novel) sequences.
2. Mapping: assembling reads against an existing backbone sequence, building a sequence that is similar but not necessarily identical to the backbone sequence.

In terms of complexity and time requirements, de-novo assemblies are orders of magnitude slower and more memory intensive than mapping assemblies. This is mostly due to the fact that the assembly algorithm needs to compare every read with every other read (an operation that has a naive time complexity of $O(n^2)$; using a hash this can be reduced significantly). Referring to the comparison drawn to shredded books in the introduction: while for mapping assemblies one would have a very similar book as template (perhaps with the names of the main characters and a few locations changed), the de-novo assemblies are more hardcore in a sense as one would not know beforehand whether this would become a science book, a novel, a catalogue, or even several books. Also, every shred would be compared with every other shred.

Influence of Technological Changes

The complexity of sequence assembly is driven by two major factors: the number of fragments and their lengths. While more and longer fragments allow better identification of sequence overlaps, they also pose problems as the underlying algorithms show quadratic or even exponential complexity behaviour to both number of fragments and their length. And while shorter sequences are faster to align, they also complicate the layout phase of an assembly as shorter reads are more difficult to use with repeats or near identical repeats.

In the earliest days of DNA sequencing, scientists could only gain a few sequences of short length (some dozen bases) after weeks of work in laboratories. Hence, these sequences could be aligned in a few minutes by hand.

In 1975, the *Dideoxy termination* method (AKA *Sanger sequencing*) was invented and until shortly after 2000, the technology was improved up to a point where fully automated machines could churn out sequences in a highly parallelised mode 24 hours a day. Large genome centers around the world housed complete farms of these sequencing machines, which in turn led to the necessity of assemblers to be optimised for sequences from whole-genome shotgun sequencing projects where the reads.

- Are about 800–900 bases long.
- Contain sequencing artifacts like sequencing and cloning vectors.
- Have error rates between 0.5 and 10%.

With the Sanger technology, bacterial projects with 20,000 to 200,000 reads could easily be assembled on one computer. Larger projects, like the human genome with approximately 35 million reads, needed large computing farms and distributed computing.

By 2004 / 2005, pyrosequencing had been brought to commercial viability by 454 Life Sciences. This new sequencing method generated reads much shorter than those of Sanger sequencing: initially about 100 bases, now 400-500 bases. Its much higher throughput and lower cost (compared to Sanger sequencing) pushed the adoption of this technology by genome centers, which in turn pushed development of sequence assemblers that could efficiently handle the read sets. The sheer amount of data coupled with technology-specific error patterns in the reads delayed development of assemblers; at the beginning in 2004 only the Newbler assembler from 454 was available. Released in mid-2007, the hybrid version of the MIRA assembler by Chevreux et al. was the first freely available assembler that could assemble 454 reads as well as mixtures of 454 reads and Sanger reads. Assembling sequences from different sequencing technologies was subsequently coined *hybrid assembly*.

From 2006, the Illumina (previously Solexa) technology has been available and can generate about 100 million reads per run on a single sequencing machine. Compare this to the 35 million reads of the human genome project which needed several years to be produced on hundreds of sequencing machines. Illumina was initially limited to a length of only 36 bases, making it less suitable for de novo assembly (such as de novo transcriptome assembly), but newer iterations of the technology achieve read lengths above 100 bases from both ends of a 3-400bp clone. Announced at the end of 2007, the SHARCGS assembler by Dohm et al. was the first published assembler that was used for an assembly with Solexa reads. It was quickly followed by a number of others.

Later, new technologies like SOLiD from Applied Biosystems, Ion Torrent and SMRT were released and new technologies (e.g. Nanopore sequencing) continue to emerge.

Greedy Algorithm

Given a set of sequence fragments the object is to find the shortest common supersequence:

1. Calculate pairwise alignments of all fragments.

2. Choose two fragments with the largest overlap.
3. Merge chosen fragments.
4. Repeat step 2 and 3 until only one fragment is left.

The result need not be an optimal solution to the problem.

Available Assemblers

The following table lists assemblers that have a de-novo assembly capability on at least one of the supported technologies.

Name	Type	Technologies	Author	Presented / Last up-dated	Licence*
ABySS	(large) genomes	Solexa, SOLiD	Simpson, J. et al.	2008 / 2014	NC-A
ALLPATHS-LG	(large) genomes	Solexa, SOLiD	Gnerre, S. et al.	2011	OS
AMOS	genomes	Sanger, 454	Salzberg, S. et al.	2002? / 2011	OS
Arapan-M	Medium Genomes (e.g. E.coli)	All	Sahli, M. & Shibuya, T.	2011 / 2012	OS
Arapan-S	Small Genomes (Viruses and Bacteria)	All	Sahli, M. & Shibuya, T.	2011 / 2012	OS
Celera WGA Assembler / CABOG	(large) genomes	Sanger, 454, Solexa	Myers, G. et al.; Miller G. et al.	2004 / 2015	OS
CLC Genomics Workbench & CLC Assembly Cell	genomes	Sanger, 454, Solexa, SOLiD	CLC bio	2008 / 2010 / 2014	C
Cortex	genomes	Solexa, SOLiD	Iqbal, Z. et al.	2011	OS
DBG2OLC	(large) genomes	Illumina, PacBio, Oxford Nanopore	Ye, C. et al	2014/2016	OS
DNA Baser Assembler	(small) genomes	Sanger, 454	Heracle BioSoft SRL	04.2016	C
DNA Dragon	genomes	Illumina, SOLiD, Complete Genomics, 454, Sanger	SequentiX	2011	C
DNAnexus	genomes	Illumina, SOLiD, Complete Genomics	DNAnexus	2011	C

DNASTAR Lasergene Genomics Suite	(large) genomes, exomes, transcriptomes, metagenomes, ESTs	Illumina, ABI SOLiD, Roche 454, Ion Torrent, Solexa, Sanger	DNASTAR	2007 / 2016	C
Edena	genomes	Illumina	D. Hernandez, P. François, L. Farinelli, M. Osteras, and J. Schrenzel.	2008/2013	OS
Euler	genomes	Sanger, 454 (‚Solexa ?)	Pevzner, P. et al.	2001 / 2006?	(C / NC-A?)
Euler-sr	genomes	454, Solexa	Chaisson, MJ. et al.	2008	NC-A
Fermi	(large) genomes	Illumina	Li, H.	2012	OS
Forge	(large) genomes, EST, metagenomes	454, Solexa, SOLID, Sanger	Platt, DM, Evers, D.	2010	OS
Geneious	genomes	Sanger, 454, Solexa, Ion Torrent, Complete Genomics, PacBio, Oxford Nanopore, Illumina	Biomatters Ltd	2009 / 2013	C
Graph Constructor	(large) genomes	Sanger, 454, Solexa, SOLiD	Convey Computer Corporation	2011	C
HINGE	genomes	PacBio/Oxford Nanopore	Kamath, Shomorony, Xia et. al.	2016	OS
IDBA (Iterative De Bruijn graph short read Assembler)	(large) genomes	Sanger,454,Solexa	Yu Peng, Henry C. M. Leung, Siu-Ming Yiu, Francis Y. L. Chin	2010	(C / NC-A?)
LIGR Assembler (derived from TIGR Assembler)	genomic	Sanger	-	2009/ 2012	OS
MaSuRCA (Maryland Super Read - Celera Assembler)	(large) genomes	Sanger, Illumina, 454	Aleksey Zimin, Guillaume Marçais, Daniela Puiu, Michael Roberts, Steven L. Salzberg, James A. Yorke	2012 / 2013	OS

MIRA (Mimicking Intelligent Read Assembly)	genomes, ESTs	Sanger, 454, Solexa	Chevreux, B.	1998 / 2014	OS
NextGENe	(small genomes?)	454, Solexa, SOLiD	Softgenetics	2008	C
Newbler	genomes, ESTs	454, Sanger	454/Roche	2004/2012	C
PADENA	genomes	454, Sanger	454/Roche	2010	OS
PASHA	(large) genomes	Illumina	Liu, Schmidt, Maskell	2011	OS
Phrap	genomes	Sanger, 454, Solexa	Green, P.	1994 / 2008	C / NC-A
TIGR Assembler	genomic	Sanger	-	1995 / 2003	OS
Trinity	Transcriptomes	short reads (paired, oriented, mixed) Illumina, 454, Solid,...	Grabher, MG et al.	2011/2016	OS
Ray	genomes	Illumina, mix of Illumina and 454, paired or not	Sébastien Boisvert, François Laviolette & Jacques Corbeil.	2010	OS [GNU General Public License]
Sequencher	genomes	traditional and next generation sequence data	Gene Codes Corporation	1991 / 2009 / 2011	C
SGA	(large) genomes	Illumina, Sanger (Roche 454?, Ion Torrent?)	Simpson, J.T. et al.	2011 / 2012	OS
SHARCGS	(small) genomes	Solexa	Dohm et al.	2007 / 2007	OS
SOPRA	genomes	Illumina, SOLiD, Sanger, 454	Dayarian, A. et al.	2010 / 2011	OS
SparseAssembler	(large) genomes	Illumina, 454, Ion torrent	Ye, C. et al.	2012 / 2012	OS
SSAKE	(small) genomes	Solexa (SOLiD? Helicos?)	Warren, R. et al.	2007 / 2014	OS
SOAPdenovo	genomes	Solexa	Luo, R. et al.	2009 / 2013	OS
SPAdes	(small) genomes, single-cell	Illumina, Solexa, Sanger, 454, Ion Torrent, PacBio, Oxford Nanopore	Bankevich, A et al.	2012 / 2015	OS
Staden gap4 package	BACs (, small genomes?)	Sanger	Staden et al.	1991 / 2008	OS
Taipan	(small) genomes	Illumina	Schmidt, B. et al.	2009 / 2009	OS

VCAKE	(small) genomes	Solexa (SOLiD?, Helicos?)	Jeck, W. et al.	2007 / 2009	OS
Phusion assembler	(large) genomes	Sanger	Mullikin JC, et al.	2003 / 2006	OS
Quality Value Guided SRA (QSRA)	genomes	Sanger, Solexa	Bryant DW, et al.	2009 / 2009	OS
Velvet	(small) genomes	Sanger, 454, Solexa, SOLiD	Zerbino, D. et al.	2007 / 2011	OS
*Licences: OS = Open Source; C = Commercial; C / NC-A = Commercial but free for non-commercial and academics; Brackets = unclear, but most likely C / NC-A					

Alignment-free Sequence Analysis

On bioinformatics, alignment-free sequence analysis approaches to molecular sequence and structure data provide alternatives over alignment-based approaches.

The emergence and need for the analysis of different types of data generated through biological research has given rise to the field of bioinformatics. Molecular sequence and structure data of DNA, RNA, and proteins, gene expression profiles or microarray data, metabolic pathway data are some of the major types of data being analysed in bioinformatics. Among them sequence data is increasing at the exponential rate due to advent of next-generation sequencing technologies. Since the origin of bioinformatics, sequence analysis has remained the major area of research with wide range of applications in database searching, genome annotation, comparative genomics, molecular phylogeny and gene prediction. The pioneering approaches for sequence analysis were based on sequence alignment either global or local, pairwise or multiple sequence alignment. Alignment-based approaches generally give excellent results when the sequences under study are closely related and can be reliably aligned, but when the sequences are divergent, a reliable alignment cannot be obtained and hence the applications of sequence alignment are limited. Another limitation of alignment-based approaches is their computational complexity and are time-consuming and thus, are limited when dealing with large-scale sequence data. The advent of next-generation sequencing technologies has resulted in generation of voluminous sequencing data. The size of this sequence data poses challenges on alignment-based algorithms in their assembly, annotation and comparative studies.

Alignment-free Methods

Alignment-free methods can broadly be classified into four categories: a) methods based on *k*-mer/word frequency, b) methods based on substrings, c) methods based on information theory and d) methods based on graphical representation. Alignment-free approaches have been used in sequence similarity searches, clustering and classification of sequences, and more recently in phylogenetics (Figure 1).

Such molecular phylogeny analyses employing alignment-free approaches are said to be part of *next-generation phylogenomics*. A number of review articles provide in-depth review of alignment-free methods in sequence analysis.

Methods Based on *k*-mer/Word Frequency

The popular methods based on *k*-mer/word frequencies include feature frequency profile (FFP), Composition vector (CV), Return time distribution (RTD), frequency chaos game representation (FCGR) and Spaced Words.

Feature Frequency Profile (Ffp)

The methodology involved in FFP based method starts by calculating the count of each possible *k*-mer (possible number of *k*-mers for nucleotide sequence: 4^k, while that for protein sequence: 20^k) in sequences. Each *k*-mer count in each sequence is then normalized by dividing it by total of all *k*-mers' count in that sequence. This leads to conversion of each sequence into its feature frequency profile. The pair wise distance between two sequences is then calculated Jensen–Shannon (JS) divergence between their respective FFPs. The distance matrix thus obtained can be used to construct phylogenetic tree using clustering algorithms like neighbor-joining, UPGMA etc.

Composition Vector (CV)

In this method frequency of appearance of each possible *k*-mer in a given sequence is calculated. The next characteristic step of this method is the subtraction of random background of these frequencies using Markov model to reduce the influence of random neutral mutations to highlight the role of selective evolution. The normalized frequencies are put a fixed order to form the composition vector (CV) of a given sequence. Cosine distance function is then used to compute pairwise distance between CVs of sequences. The distance matrix thus obtained can be used to construct phylogenetic tree using clustering algorithms like neighbor-joining, UPGMA etc. This method can be extended through resort to efficient pattern matching algorithms to include in the computation of the composition vectors: (i) all *k*-mers for any value of *k*, (ii) all substrings of any length up to an arbitrarily set maximum *k* value, (iii) all maximal substrings, where a substring is maximal if extending it by any character would cause a decrease in its occurrence count.

Return Time Distribution (RTD)

The RTD based method does not calculate the count of k-mers in sequences, instead it computes the time required for the reappearance of k-mers. The time refers to the number of residues in successive appearance of particular k-mer. Thus the occurrence of each k-mer in a sequence is calculated in the form of RTD, which is then summarised using two statistical parameters mean (μ) and standard deviation (σ). Thus each sequence is represented in the form of numeric vector of size $2 \cdot 4^k$ containing μ and σ of 4^k RTDs. The pair wise distance between sequences is calculated using Euclidean distance measure. The distance matrix thus obtained can be used to construct phylogenetic tree using clustering algorithms like neighbor-joining, UPGMA etc.

Frequency Chaos Game Representation (FCGR)

The FCGR methods have evolved from chaos game representation (CGR) technique, which provides scale independent representation for genomic sequences. The CGRs can be divided by grid lines where each grid square denotes the occurrence of oligonucleotides of a specific length in the sequence. Such representation of CGRs is termed as Frequency Chaos Game Representation (FCGR). This leads to representation of each sequence into FCGR. The pair wise distance between FCGRs of sequences can be calculated using the Pearson distance, the Hamming distance or the Euclidean distance.

Spaced-word Frequencies

While most alignment-free algorithms compare the word-composition of sequences, Spaced Words uses a pattern of care and don't care positions. The occurrence of a spaced word in a sequence is then defined by the characters at the match positions only, while the characters at the don't care positions are ignored. Instead of comparing the frequencies of contiguous words in the input sequences, this approach compares the frequencies of the spaced words according to the pre-defined pattern.

Methods Based on Substrings

The methods in this category employ the similarity and differences of substrings in a pair of sequences. These algorithms were mostly used for string processing in computer science.

Average Common Substring (ACS)

In this approach, for a chosen pair of sequences (A and B of lengths ℓ and m respectively), longest substring starting at some position is identified in one sequence (A) which exactly matches in the other sequence (B) at any position. In this way, lengths of longest substrings starting at different positions in sequence A and having exact matches at some positions in sequence B are calculated. All these lengths are averaged to derive a measure $L(A,B)$. Intuitively, larger the $L(A,B)$, the more similar the two sequences are. To account for the differences in the length of sequences, $L(A,B)$ is normalized [i.e. $L(A,B)>$]. This gives the similarity measure between the sequences.

In order to derive a distance measure, the inverse of similarity measure is taken and a correction term is subtracted from it to assure that $d(A,A)$ will be zero. Thus:

$$d(A,B)=\left[\frac{\log m}{L(A,B)}\right]-\left[\frac{\log n}{L(A,A)}\right].$$

This measure $d(A,B)$ is not symmetric, so one has to compute $d_s(A,B)=d_s(B,A)=(d(A,B)+d(B,A))/2$, which gives final ACS measure between the two strings (A and B). The subsequence/substring search can be efficiently performed by using suffix trees.

K-Mismatch Average Common Substring Approach (kmacs)

This approach is a generalization of the ACS approach. To define the distance between two DNA or protein sequences, kmacs estimates for each position *i* of the first sequence the longest substring starting at *i* and matching a substring of the second sequence with up to *k* mismatches. It defines the average of these values as a measure of similarity between the sequences and turns this into a symmetric distance measure. Kmacs does not compute exact *k*-mismatch substrings, since this would be computational too costly, but approximates such substrings.

Mutation Distances (Kr)

This approach is closely related to the ACS, which calculates the number of substitutions per site between two DNA sequences using the shortest absent substring (termed as shustring).

Methods Based on Information Theory

Information Theory has provided successful methods for alignment-free sequence analysis and comparison. The existing applications of information theory include global and local characterization of DNA, RNA and proteins, estimating genome entropy to motif and region classification. It also holds promise in gene mapping, next-generation sequencing analysis and metagenomics.

Base–base Correlation (BBC)

Base–base correlation (BBC) converts the genome sequence into a unique 16-dimensional numeric vector using the following equation.

$$T_{ij}(K) = \sum_{\ell=1}^{K} P_{ij}(\ell) \cdot \log_2 \left(\frac{P_{ij}(\ell)}{P_i P_j} \right)$$

The P_i and P_j denotes the probabilities of bases *i* and *j* in the genome. The $P_{ij}(\ell)$ indicates the probability of bases *i* and *j* at distance ℓ in the genome. The parameter *K* indicates the maximum distance between the bases *i* and *j*. The variation in the values of 16 parameters reflect variation in the genome content and length.

Information Correlation and Partial Information Correlation (IC-PIC)

IC-PIC (information correlation and partial information correlation) based method employs the base correlation property of DNA sequence. IC and PIC were calculated using following formulas,

$$IC_\ell = -2\sum_i P_i \log_2 P_i + \sum_{ij} P_{ij}(\ell) \log_2 P_{ij}(\ell)$$

$$PIC_{ij}(\ell) = (P_{ij}(\ell) - P_i P_j(\ell))^2$$

The final vector is obtained as follows:

$$V = \frac{IC_\ell}{PIC_{ij}(\ell)} \text{ where } \ell \in \{\ell_0, \ell_0 + 1, \ldots, \ell_0 + n\},$$

which defines the range of distance between bases.

The pair wise distance between sequences is calculated using Euclidean distance measure. The distance matrix thus obtained can be used to construct phylogenetic tree using clustering algorithms like neighbor-joining, UPGMA etc.

Lempel–Ziv Compress

Lempel–Ziv complexity uses the relative information between the sequences. This complexity is measured by the number of steps required to generate a string given the prior knowledge of another string and a self-delimiting production process. This measure has a relation to measuring *k*-words in a sequence, as they can be easily used to generate the sequence. It is computational intensive method. Otu and Sayood (2003) used this method to construct five different distance measures for phylogenetic tree construction.

Context Modeling Compress

In the context modeling complexity the next-symbol predictions, of one or more statistical models, are combined or competing to yield a prediction that is based on events recorded in the past. The algorithmic information content derived from each symbol prediction can be used to compute algorithmic information profiles with a time proportional to the length of the sequence. The process has been applied to DNA sequence analysis.

Methods Based on Graphical Representation

Iterated Maps

The use of iterated maps for sequence analysis was first introduced by HJ Jefferey in 1990 when he proposed to apply the Chaos Game to map genomic sequences into a unit square. That report coined the procedure as Chaos Game Representation (CGR). However, only 3 years later this approach was first dismissed as a projection of a Markov transition table by N Goldman. This objection was overruled by the end of that decade when the opposite was found to be the case – that CGR bijectively maps Markov transition is into a fractal, order-free (degree-free) representation. The realization that iterated maps provide a bijective map between the symbolic space and numeric space led to the identification of a variety of alignment-free approaches to sequence comparison and characterization. These developments were reviewed in late 2013 by JS Almeida in. A number of web apps such as http://usm.github.com are available to demonstrate how to encode and compare arbitrary symbolic sequences.

Comparison of Alignment Based and Alignment-free Methods

Alignment-based methods	Alignment-free methods
These methods assume that homologous regions are contiguous (with gaps).	Does not assume such contiguity of homologous regions.
Computes all possible pairwise comparisons of sequences; hence computationally expensive.	Based on occurrences of sub-sequences; composition; computationally inexpensive, can be memory-intensive.
Well-established approach in phylogenomics.	Relatively recent and application in phylogenomics is limited; needs further testing for robustness and scalability.
Requires substitution/evolutionary models.	Less dependent on substitution/evolutionary models.
Sensitive to stochastic sequence variation, recombination, horizontal (or lateral) genetic transfer, rate heterogeneity and sequences of varied lengths, especially when similarity lies in the "twilight zone".	Less sensitive to stochastic sequence variation, recombination, horizontal (or lateral) genetic transfer, rate heterogeneity and sequences of varied lengths.
Best practice uses inference algorithms with complexity at least $O(n^2)$; less time-efficient.	Inference algorithms typically $O(n^2)$ or less; more time-efficient.

Heuristic in nature; statistical significance of how alignment scores relate to homology is difficult to assess.	Exact solutions; statistical significance of the sequence distances (and degree of similarity) can be readily assessed.

Applications of Alignment-free Methods

- Genomic rearrangements,
- Molecular phylogenetics,
- Metagenomics,
- Next generation sequence data analysis,
- Epigenomics,
- Barcoding of species,
- Population genetics,
- Horizontal gene transfer,
- Sero/genotyping of viruses,
- Allergenicity prediction,
- SNP discovery,
- Recombination detection.

List of Web Servers/Software for Alignment-free Methods

Name	Description	Availability
kmacs	k-mismatch average common substring approach.	kmacs
Spaced words	Spaced-word frequencies.	spaced-words
FFP	Feature frequency profile based phylogeny.	FFP
CVTree	Composition vector based server for phylogeny.	CVTree
RTD Phylogeny	Return time distribution based server for phylogeny.	RTD Phylogeny
AGP	A multimethods web server for alignment-free genome phylogeny.	AGP
Alfy	Alignment-free detection of local similarity among viral and bacterial genomes.	Alfy
decaf+py	DistancE Calculation using Alignment-Free methods in PYthon.	decaf+py
Dengue Subtyper	Genotyping of Dengue viruses based on RTD.	Dengue Subtyper
WNV Typer	Genotyping of West nile viruses based on RTD.	WNV Typer

AllergenFP	Allergenicity prediction by descriptor fingerprints.	AllergenFP
kSNP v2	Alignment-Free SNP Discovery.	kSNP v2
d2Tools	Comparison of Metatranscriptomic Samples Based on k-Tuple Frequencies.	d2Tools
rush	Recombination detection Using SHustrings.	rush
smash	Genomic rearrangements detection and visualisation.	smash
GScompare	Oligonucleotide-based fast clustering of bacterial genomes.	GScompare
COMET	Alignment-free subtyping of HIV-1, HIV-2 and HCV viral sequences.	COMET

Sequence Alignment

In bioinformatics, a sequence alignment is a way of arranging the sequences of DNA, RNA, or protein to identify regions of similarity that may be a consequence of functional, structural, or evolutionary relationships between the sequences. Aligned sequences of nucleotide or amino acid residues are typically represented as rows within a matrix. Gaps are inserted between the residues so that identical or similar characters are aligned in successive columns. Sequence alignments are also used for non-biological sequences, such as calculating the edit distance cost between strings in a natural language or in financial data.

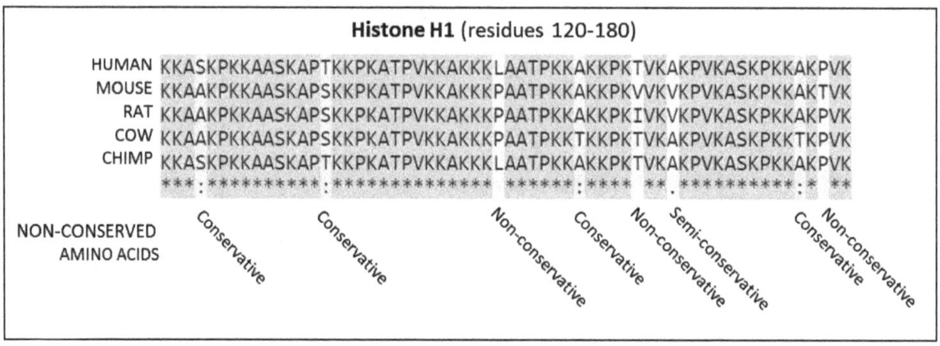

A sequence alignment, produced by ClustalO, of mammalian histone proteins. Sequences are the amino acids for residues 120-180 of the proteins. Residues that are conserved across all sequences are highlighted in grey. Below the protein sequences is a key denoting conserved sequence (*), conservative mutations (:), semi-conservative mutations (.), and non-conservative mutations ( ).

Interpretation

If two sequences in an alignment share a common ancestor, mismatches can be interpreted as point mutations and gaps as indels (that is, insertion or deletion mutations) introduced in one or both lineages in the time since they diverged from one another.

In sequence alignments of proteins, the degree of similarity between amino acids occupying a particular position in the sequence can be interpreted as a rough measure of how conserved a particular region or sequence motif is among lineages. The absence of substitutions, or the presence of only very conservative substitutions (that is, the substitution of amino acids whose side chains have similar biochemical properties) in a particular region of the sequence, suggest that this region has structural or functional importance. Although DNA and RNA nucleotide bases are more similar to each other than are amino acids, the conservation of base pairs can indicate a similar functional or structural role.

Alignment Methods

Very short or very similar sequences can be aligned by hand. However, most interesting problems require the alignment of lengthy, highly variable or extremely numerous sequences that cannot be aligned solely by human effort. Instead, human knowledge is applied in constructing algorithms to produce high-quality sequence alignments, and occasionally in adjusting the final results to reflect patterns that are difficult to represent algorithmically (especially in the case of nucleotide sequences). Computational approaches to sequence alignment generally fall into two categories: *global alignments* and *local alignments*. Calculating a global alignment is a form of global optimization that "forces" the alignment to span the entire length of all query sequences. By contrast, local alignments identify regions of similarity within long sequences that are often widely divergent overall. Local alignments are often preferable, but can be more difficult to calculate because of the additional challenge of identifying the regions of similarity. A variety of computational algorithms have been applied to the sequence alignment problem. These include slow but formally correct methods like dynamic programming. These also include efficient, heuristic algorithms or probabilistic methods designed for large-scale database search, that do not guarantee to find best matches.

Representations

Alignments are commonly represented both graphically and in text format. In almost all sequence alignment representations, sequences are written in rows arranged so that aligned residues appear in successive columns. In text formats, aligned columns containing identical or similar characters are indicated with a system of conservation symbols. As in the image above, an asterisk or pipe symbol is used to show identity between two columns; other less common symbols include a colon for conservative substitutions and a period for semiconservative substitutions. Many sequence visualization programs also use color to display information about the properties of the individual sequence elements; in DNA and RNA sequences, this equates to assigning each nucleotide its own color. In protein alignments, such as the one in the image above, color is often used to indicate amino acid properties to aid in judging the conservation of a given amino acid substitution. For multiple sequences the last row in each column is often the consensus sequence determined by the alignment; the consensus sequence is also often represented in graphical format with a sequence logo in which the size of each nucleotide or amino acid letter corresponds to its degree of conservation.

Sequence alignments can be stored in a wide variety of text-based file formats, many of which were originally developed in conjunction with a specific alignment program or implementation. Most web-based tools allow a limited number of input and output formats, such as FASTA format and GenBank format and the output is not easily editable. Several conversion programs that provide graphical and/or command line interfaces are available, such as READSEQ and EMBOSS. There are also several programming packages which provide this conversion functionality, such as BioPython, BioRuby and BioPerl.

Global and Local Alignments

Global alignments, which attempt to align every residue in every sequence, are most useful when the sequences in the query set are similar and of roughly equal size. (This does not mean global alignments cannot start and/or end in gaps.) A general global alignment technique is the Needleman–Wunsch algorithm, which is based on dynamic programming. Local alignments are more useful for dissimilar sequences that are suspected to contain regions of similarity or similar sequence motifs within their larger sequence context. The Smith–Waterman algorithm is a general local alignment method also based on dynamic programming.

Hybrid methods, known as semiglobal or "glocal" (short for global-local) methods, attempt to find the best possible alignment that includes the start and end of one or the other sequence. This can be especially useful when the downstream part of one sequence overlaps with the upstream part of the other sequence. In this case, neither global nor local alignment is entirely appropriate: a global alignment would attempt to force the alignment to extend beyond the region of overlap, while a local alignment might not fully cover the region of overlap. Another case where semiglobal alignment is useful is when one sequence is short (for example a gene sequence) and the other is very long (for example a chromosome sequence). In that case, the short sequence should be globally aligned but only a local alignment is desired for the long sequence.

Pairwise Alignment

Pairwise sequence alignment methods are used to find the best-matching piecewise (local) or global alignments of two query sequences. Pairwise alignments can only be used between two sequences at a time, but they are efficient to calculate and are often used for methods that do not require extreme precision (such as searching a database for sequences with high similarity to a query). The three primary methods of producing pairwise alignments are dot-matrix methods, dynamic programming, and word methods; however, multiple sequence alignment techniques can also align pairs of sequences. Although each method has its individual strengths and weaknesses, all three pairwise methods have difficulty with highly repetitive sequences of low information content - especially where the number of repetitions differ in the two sequences to be aligned. One way of quantifying the utility of a given pairwise alignment is the 'maximum unique match' (MUM), or the longest subsequence that occurs in both query sequences. Longer MUM sequences typically reflect closer relatedness.

Dot-matrix Methods

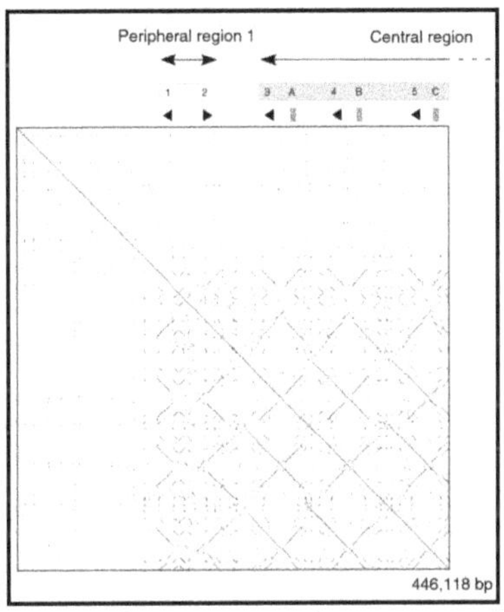

Self comparison of a part of a mouse strain genome. The dot-plot shows a patchwork of lines, demonstrating duplicated segments of DNA.

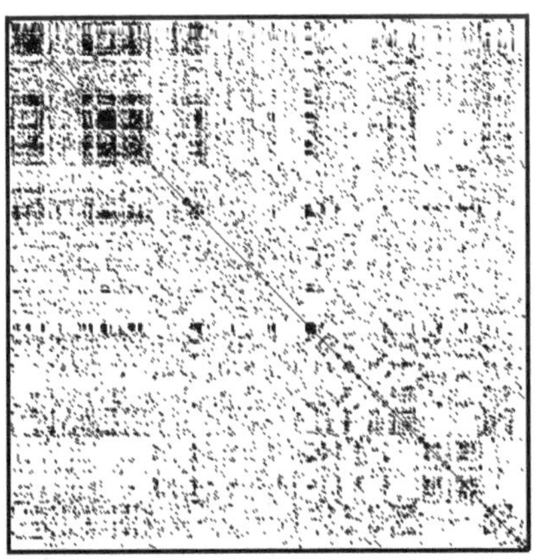

A DNA dot plot of a human zinc finger transcription factor (GenBank ID NM_002383), showing regional self-similarity. The main diagonal represents the sequence's alignment with itself; lines off the main diagonal represent similar or repetitive patterns within the sequence. This is a typical example of a recurrence plot.

The dot-matrix approach, which implicitly produces a family of alignments for individual sequence regions, is qualitative and conceptually simple, though time-consuming to analyze on a large scale. In the absence of noise, it can be easy to visually identify certain sequence features—such as insertions, deletions, repeats, or inverted repeats—from a dot-matrix plot. To construct a dot-matrix plot, the two sequences are written along the top row and leftmost column of a two-dimensional matrix and a dot is placed at any point where the characters in the appropriate columns match—this is a typical recurrence plot. Some implementations vary the size or intensity of the dot depending on the degree of similarity of the two characters, to accommodate conservative substitutions. The dot plots of very closely related sequences will appear as a single line along the matrix's main diagonal.

Problems with dot plots as an information display technique include: noise, lack of clarity, non-intuitiveness, difficulty extracting match summary statistics and match positions on the two sequences. There is also much wasted space where the match data is inherently duplicated across the diagonal and most of the actual area of the plot is taken up by either empty space or noise, and, finally, dot-plots are limited to two sequences. None of these limitations apply to Miropeats alignment diagrams but they have their own particular flaws.

Dot plots can also be used to assess repetitiveness in a single sequence. A sequence can be plotted against itself and regions that share significant similarities will appear as lines off the main diagonal. This effect can occur when a protein consists of multiple similar structural domains.

Dynamic Programming

The technique of dynamic programming can be applied to produce global alignments via the Needleman-Wunsch algorithm, and local alignments via the Smith-Waterman algorithm. In typical usage, protein alignments use a substitution matrix to assign scores to amino-acid matches or mismatches, and a gap penalty for matching an amino acid in one sequence to a gap in the other. DNA and RNA alignments may use a scoring matrix, but in practice often simply assign a positive match score, a negative mismatch score, and a negative gap penalty. (In standard dynamic programming, the score of each amino acid position is independent of the identity of its neighbors, and therefore base stacking effects are not taken into account. However, it is possible to account for such effects by modifying the algorithm.) A common extension to standard linear gap costs, is the usage of two different gap penalties for opening a gap and for extending a gap. Typically the former is much larger than the latter, e.g. -10 for gap open and -2 for gap extension. Thus, the number of gaps in an alignment is usually reduced and residues and gaps are kept together, which typically makes more biological sense. The Gotoh algorithm implements affine gap costs by using three matrices.

Dynamic programming can be useful in aligning nucleotide to protein sequences, a task complicated by the need to take into account frameshift mutations (usually insertions or deletions). The framesearch method produces a series of global or local pairwise alignments between a query nucleotide sequence and a search set of protein sequences, or vice versa. Its ability to evaluate frameshifts offset by an arbitrary number of nucleotides makes the method useful for sequences containing large numbers of indels, which can be very difficult to align with more efficient heuristic methods. In practice, the method requires large amounts of computing power or a system whose architecture is specialized for dynamic programming. The BLAST and EMBOSS suites provide basic tools for creating translated alignments (though some of these approaches take advantage of side-effects of sequence searching capabilities of the tools). More general methods are available from both commercial sources, such as *FrameSearch*, distributed as part of the Accelrys GCG package, and Open Source software such as Genewise.

The dynamic programming method is guaranteed to find an optimal alignment given a particular scoring function; however, identifying a good scoring function is often an empirical rather than a theoretical matter. Although dynamic programming is extensible to more than two sequences, it is prohibitively slow for large numbers of sequences or extremely long sequences.

Word Methods

Word methods, also known as *k*-tuple methods, are heuristic methods that are not guaranteed to find an optimal alignment solution, but are significantly more efficient than dynamic programming. These methods are especially useful in large-scale database searches where it is understood that a large proportion of the candidate sequences will have essentially no significant match with the query sequence. Word methods are best known for their implementation in the database search tools FASTA and the BLAST family. Word methods identify a series of short, nonoverlapping subsequences ("words") in the query sequence that are then matched to candidate database sequences. The relative positions of the word in the two sequences being compared are subtracted to obtain an offset; this will indicate a region of alignment if multiple distinct words produce the same offset. Only if this region is detected do these methods apply more sensitive alignment criteria; thus, many unnecessary comparisons with sequences of no appreciable similarity are eliminated.

In the FASTA method, the user defines a value *k* to use as the word length with which to search the database. The method is slower but more sensitive at lower values of *k*, which are also preferred for searches involving a very short query sequence. The BLAST family of search methods provides a number of algorithms optimized for particular types of queries, such as searching for distantly related sequence matches. BLAST was developed to provide a faster alternative to FASTA without sacrificing much accuracy; like FASTA, BLAST uses a word search of length *k*, but evaluates only the most significant word matches, rather than every word match as does FASTA. Most BLAST implementations use a fixed default word length that is optimized for the query and database type, and that is changed only under special circumstances, such as when searching with repetitive or very short query sequences. Implementations can be found via a number of web portals, such as EMBL FASTA and NCBI BLAST.

Multiple Sequence Alignment

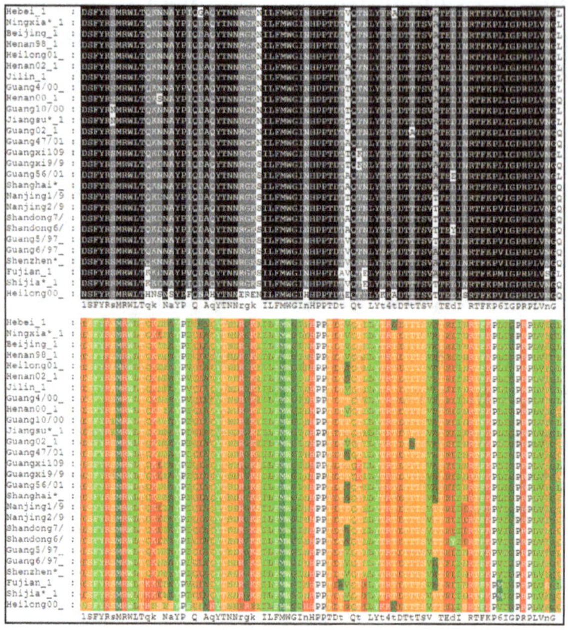

Alignment of 27 avian influenza hemagglutinin protein sequences colored by residue conservation (top) and residue properties (bottom).

Multiple sequence alignment is an extension of pairwise alignment to incorporate more than two sequences at a time. Multiple alignment methods try to align all of the sequences in a given query set. Multiple alignments are often used in identifying conserved sequence regions across a group of sequences hypothesized to be evolutionarily related. Such conserved sequence motifs can be used in conjunction with structural and mechanistic information to locate the catalytic active sites of enzymes. Alignments are also used to aid in establishing evolutionary relationships by constructing phylogenetic trees. Multiple sequence alignments are computationally difficult to produce and most formulations of the problem lead to NP-complete combinatorial optimization problems. Nevertheless, the utility of these alignments in bioinformatics has led to the development of a variety of methods suitable for aligning three or more sequences.

Dynamic Programming

The technique of dynamic programming is theoretically applicable to any number of sequences; however, because it is computationally expensive in both time and memory, it is rarely used for more than three or four sequences in its most basic form. This method requires constructing the n-dimensional equivalent of the sequence matrix formed from two sequences, where n is the number of sequences in the query. Standard dynamic programming is first used on all pairs of query sequences and then the "alignment space" is filled in by considering possible matches or gaps at intermediate positions, eventually constructing an alignment essentially between each two-sequence alignment. Although this technique is computationally expensive, its guarantee of a global optimum solution is useful in cases where only a few sequences need to be aligned accurately. One method for reducing the computational demands of dynamic programming, which relies on the "sum of pairs" objective function, has been implemented in the MSA software package.

Progressive Methods

Progressive, hierarchical, or tree methods generate a multiple sequence alignment by first aligning the most similar sequences and then adding successively less related sequences or groups to the alignment until the entire query set has been incorporated into the solution. The initial tree describing the sequence relatedness is based on pairwise comparisons that may include heuristic pairwise alignment methods similar to FASTA. Progressive alignment results are dependent on the choice of "most related" sequences and thus can be sensitive to inaccuracies in the initial pairwise alignments. Most progressive multiple sequence alignment methods additionally weight the sequences in the query set according to their relatedness, which reduces the likelihood of making a poor choice of initial sequences and thus improves alignment accuracy.

Many variations of the Clustal progressive implementation are used for multiple sequence alignment, phylogenetic tree construction, and as input for protein structure prediction. A slower but more accurate variant of the progressive method is known as T-Coffee.

Iterative Methods

Iterative methods attempt to improve on the heavy dependence on the accuracy of the initial pairwise alignments, which is the weak point of the progressive methods. Iterative methods optimize an objective function based on a selected alignment scoring method by assigning an initial global alignment and then realigning sequence subsets. The realigned subsets are then themselves aligned to produce the next iteration's multiple sequence alignment. Various ways of selecting the sequence subgroups and objective function are reviewed in.

Motif Finding

Motif finding, also known as profile analysis, constructs global multiple sequence alignments that attempt to align short conserved sequence motifs among the sequences in the query set. This is usually done by first constructing a general global multiple sequence alignment, after which the highly conserved regions are isolated and used to construct a set of profile matrices. The profile matrix for each conserved region is arranged like a scoring matrix but its frequency counts for each amino acid or nucleotide at each position are derived from the conserved region's character distribution rather than from a more general empirical distribution. The profile matrices are then used to search other sequences for occurrences of the motif they characterize. In cases where the original data set contained a small number of sequences, or only highly related sequences, pseudocounts are added to normalize the character distributions represented in the motif.

Techniques Inspired by Computer Science

A variety of general optimization algorithms commonly used in computer science have also been applied to the multiple sequence alignment problem. Hidden Markov models have been used to produce probability scores for a family of possible multiple sequence alignments for a given query set; although early HMM-based methods produced underwhelming performance, later applications have found them especially effective in detecting remotely related sequences because they are less susceptible to noise created by conservative or semiconservative substitutions. Genetic algorithms and simulated annealing have also been used in optimizing multiple sequence alignment scores as judged by a scoring function like the sum-of-pairs method. More complete details and software packages can be found in the main article multiple sequence alignment.

The Burrows–Wheeler transform has been successfully applied to fast short read alignment in popular tools such as Bowtie and BWA.

Structural Alignment

Structural alignments, which are usually specific to protein and sometimes RNA sequences, use information about the secondary and tertiary structure of the protein or RNA molecule to aid in aligning the sequences. These methods can be used for two or more sequences and typically produce local alignments; however, because they depend on the availability of structural information, they can only be used for sequences whose corresponding structures are known (usually through X-ray crystallography or NMR spectroscopy). Because both protein and RNA structure is more evolutionarily conserved than sequence, structural alignments can be more reliable between sequences that are very distantly related and that have diverged so extensively that sequence comparison cannot reliably detect their similarity.

Structural alignments are used as the "gold standard" in evaluating alignments for homology-based protein structure prediction because they explicitly align regions of the protein sequence that are structurally similar rather than relying exclusively on sequence information. However, clearly structural alignments cannot be used in structure prediction because at least one sequence in the query set is the target to be modeled, for which the structure is not known. It has been shown that, given the structural alignment between a target and a template sequence, highly accurate models of the target protein sequence can be produced; a major stumbling block in homology-based structure prediction is the production of structurally accurate alignments given only sequence information.

DALI

The DALI method, or distance matrix alignment, is a fragment-based method for constructing structural alignments based on contact similarity patterns between successive hexapeptides in the query sequences. It can generate pairwise or multiple alignments and identify a query sequence's structural neighbors in the Protein Data Bank (PDB). It has been used to construct the FSSP structural alignment database (Fold classification based on Structure-Structure alignment of Proteins, or Families of Structurally Similar Proteins). A DALI webserver can be accessed at DALI and the FSSP is located at The Dali Database.

SSAP

SSAP (sequential structure alignment program) is a dynamic programming-based method of structural alignment that uses atom-to-atom vectors in structure space as comparison points. It has been extended since its original description to include multiple as well as pairwise alignments, and has been used in the construction of the CATH (Class, Architecture, Topology, Homology) hierarchical database classification of protein folds. The CATH database can be accessed at CATH Protein Structure Classification.

Combinatorial Extension

The combinatorial extension method of structural alignment generates a pairwise structural alignment by using local geometry to align short fragments of the two proteins being analyzed and then assembles these fragments into a larger alignment. Based on measures such as rigid-body root mean square distance, residue distances, local secondary structure, and surrounding environmental features such as residue neighbor hydrophobicity, local alignments called "aligned fragment pairs" are generated and used to build a similarity matrix representing all possible structural alignments within predefined cutoff criteria. A path from one protein structure state to the other is then traced through the matrix by extending the growing alignment one fragment at a time. The optimal such path defines the combinatorial-extension alignment. A web-based server implementing the method and providing a database of pairwise alignments of structures in the Protein Data Bank is located at the Combinatorial Extension website.

Phylogenetic Analysis

Phylogenetics and sequence alignment are closely related fields due to the shared necessity of evaluating sequence relatedness. The field of phylogenetics makes extensive use of sequence alignments in the construction and interpretation of phylogenetic trees, which are used to classify the evolutionary relationships between homologous genes represented in the genomes of divergent species. The degree to which sequences in a query set differ is qualitatively related to the sequences' evolutionary distance from one another. Roughly speaking, high sequence identity suggests that the sequences in question have a comparatively young most recent common ancestor, while low identity suggests that the divergence is more ancient. This approximation, which reflects the "molecular clock" hypothesis that a roughly constant rate of evolutionary change can be used to extrapolate the elapsed time since two genes first diverged (that is, the coalescence time), assumes that the effects of mutation and selection are constant across sequence lineages. Therefore, it does not account for possible difference among organisms or species in the rates of DNA repair or the possible functional conservation of specific regions in a sequence. (In the case of nucleotide sequences, the molecular clock hypothesis in its most basic form also discounts the difference in acceptance rates between silent mutations that do not alter the meaning of a given codon and other mutations that result in a different amino acid being incorporated into the protein.) More statistically accurate methods allow the evolutionary rate on each branch of the phylogenetic tree to vary, thus producing better estimates of coalescence times for genes.

Progressive multiple alignment techniques produce a phylogenetic tree by necessity because they incorporate sequences into the growing alignment in order of relatedness. Other techniques that assemble multiple sequence alignments and phylogenetic trees score and sort trees first and calculate a multiple sequence alignment from the highest-scoring tree. Commonly used methods of phylogenetic tree construction are mainly heuristic because the problem of selecting the optimal tree, like the problem of selecting the optimal multiple sequence alignment, is NP-hard.

Assessment of Significance

Sequence alignments are useful in bioinformatics for identifying sequence similarity, producing phylogenetic trees, and developing homology models of protein structures. However, the biological relevance of sequence alignments is not always clear. Alignments are often assumed to reflect a degree of evolutionary change between sequences descended from a common ancestor; however, it is formally possible that convergent evolution can occur to produce apparent similarity between proteins that are evolutionarily unrelated but perform similar functions and have similar structures.

In database searches such as BLAST, statistical methods can determine the likelihood of a particular alignment between sequences or sequence regions arising by chance given the size and composition of the database being searched. These values can vary significantly depending on the search space. In particular, the likelihood of finding a given alignment by chance increases if the database consists only of sequences from the same organism as the query sequence. Repetitive sequences in the database or query can also distort both the search results and the assessment of statistical significance; BLAST automatically filters such repetitive sequences in the query to avoid apparent hits that are statistical artifacts.

Methods of statistical significance estimation for gapped sequence alignments are available in the literature.

Assessment of Credibility

Statistical significance indicates the probability that an alignment of a given quality could arise by chance, but does not indicate how much superior a given alignment is to alternative alignments of the same sequences. Measures of alignment credibility indicate the extent to which the best scoring alignments for a given pair of sequences are substantially similar. Methods of alignment credibility estimation for gapped sequence alignments are available in the literature.

Scoring Functions

The choice of a scoring function that reflects biological or statistical observations about known sequences is important to producing good alignments. Protein sequences are frequently aligned using substitution matrices that reflect the probabilities of given character-to-character substitutions. A series of matrices called PAM matrices (Point Accepted Mutation matrices, originally defined by Margaret Dayhoff and sometimes referred to as "Dayhoff matrices") explicitly encode evolutionary approximations regarding the rates and probabilities of particular amino acid mutations. Another common series of scoring matrices, known as BLOSUM (Blocks Substitution Matrix), encodes empirically derived substitution probabilities. Variants of both types of matrices are used to detect sequences with differing levels of divergence, thus allowing users of BLAST or FASTA to restrict searches to more closely related matches or expand to detect more divergent sequences. Gap penalties account for the introduction of a gap - on the evolutionary model, an insertion or deletion mutation - in both nucleotide and protein sequences, and therefore the penalty values should be proportional to the expected rate of such mutations. The quality of the alignments produced therefore depends on the quality of the scoring function.

It can be very useful and instructive to try the same alignment several times with different choices for scoring matrix and/or gap penalty values and compare the results. Regions where the solution is weak or non-unique can often be identified by observing which regions of the alignment are robust to variations in alignment parameters.

Other Biological Uses

Sequenced RNA, such as expressed sequence tags and full-length mRNAs, can be aligned to a sequenced genome to find where there are genes and get information about alternative splicing and RNA editing. Sequence alignment is also a part of genome assembly, where sequences are aligned to find overlap so that *contigs* (long stretches of sequence) can be formed. Another use is SNP analysis, where sequences from different individuals are aligned to find single basepairs that are often different in a population.

Non-biological Uses

The methods used for biological sequence alignment have also found applications in other fields, most notably in natural language processing and in social sciences, where the Needleman-Wunsch algorithm is usually referred to as Optimal matching. Techniques that generate the set of elements from which words will be selected in natural-language generation algorithms have borrowed multiple sequence alignment techniques from bioinformatics to produce linguistic versions of computer-generated mathematical proofs. In the field of historical and comparative linguistics, sequence alignment has been used to partially automate the comparative method by which linguists traditionally reconstruct languages. Business and marketing research has also applied multiple sequence alignment techniques in analyzing series of purchases over time.

Software

A more complete list of available software categorized by algorithm and alignment type is available at sequence alignment software, but common software tools used for general sequence alignment tasks include ClustalW2 and T-coffee for alignment, and BLAST and FASTA3x for database searching. Commercial tools such as Geneious and PatternHunter are also available.

Alignment algorithms and software can be directly compared to one another using a standardized set of benchmark reference multiple sequence alignments known as BAliBASE. The data set consists of structural alignments, which can be considered a standard against which purely sequence-based methods are compared. The relative performance of many common alignment methods on frequently encountered alignment problems has been tabulated and selected results published online at BAliBASE. A comprehensive list of BAliBASE scores for many (currently 12) different alignment tools can be computed within the protein workbench STRAP.

BLAST

In bioinformatics, BLAST for Basic Local Alignment Search Tool is an algorithm for comparing primary biological sequence information, such as the amino-acid sequences of proteins or the nucleotides of DNA sequences. A BLAST search enables a researcher to compare a query sequence with a library or database of sequences, and identify library sequences that resemble the query sequence above a certain threshold.

Different types of BLASTs are available according to the query sequences. For example, following the discovery of a previously unknown gene in the mouse, a scientist will typically perform a BLAST search of the human genome to see if humans carry a similar gene; BLAST will identify sequences in the human genome that resemble the mouse gene based on similarity of sequence. The BLAST algorithm and program were designed by Stephen Altschul, Warren Gish, Webb Miller, Eugene Myers, and David J. Lipman at the National Institutes of Health and was published in the *Journal of Molecular Biology* in 1990 and cited over 50,000 times.

Background

BLAST is one of the most widely used bioinformatics programs for sequence searching. It addresses a fundamental problem in bioinformatics research. The heuristic algorithm it uses is much faster than other approaches, such as calculating an optimal alignment. This emphasis on speed is vital to making the algorithm practical on the huge genome databases currently available, although subsequent algorithms can be even faster.

Before BLAST, FASTA was developed by David J. Lipman and William R. Pearson in 1985.

Before fast algorithms such as BLAST and FASTA were developed, doing database searches for protein or nucleic sequences was very time consuming because a full alignment procedure (e.g., the Smith–Waterman algorithm) was used.

While BLAST is faster than any Smith-Waterman implementation for most cases, it cannot "guarantee the optimal alignments of the query and database sequences" as Smith-Waterman algorithm does. The optimality of Smith-Waterman "ensured the best performance on accuracy and the most precise results" at the expense of time and computer power.

BLAST is more time-efficient than FASTA by searching only for the more significant patterns in the sequences, yet with comparative sensitivity. This could be further realized by understanding the algorithm of BLAST introduced below.

Examples of other questions that researchers use BLAST to answer are:

- Which bacterial species have a protein that is related in lineage to a certain protein with known amino-acid sequence.
- What other genes encode proteins that exhibit structures or motifs such as ones that have just been determined.

BLAST is also often used as part of other algorithms that require approximate sequence matching.

The BLAST algorithm and the computer program that implements it were developed by Stephen Altschul, Warren Gish, and David Lipman at the U.S. National Center for Biotechnology Information (NCBI), Webb Miller at the Pennsylvania State University, and Gene Myers at the University of Arizona. It is available on the web on the NCBI website. Alternative implementations include AB-BLAST (formerly known as WU-BLAST), FSA-BLAST (last updated in 2006), and ScalaBLAST.

The original paper by Altschul, *et al.* was the most highly cited paper published in the 1990s.

Input

Input sequences (in FASTA or Genbank format) and weight matrix.

Output

BLAST output can be delivered in a variety of formats. These formats include HTML, plain text, and XML formatting. For NCBI's web-page, the default format for output is HTML. When performing a BLAST on NCBI, the results are given in a graphical format showing the hits found, a table showing sequence identifiers for the hits with scoring related data, as well as alignments for the sequence of interest and the hits received with corresponding BLAST scores for these. The easiest to read and most informative of these is probably the table.

If one is attempting to search for a proprietary sequence or simply one that is unavailable in databases available to the general public through sources such as NCBI, there is a BLAST program available for download to any computer, at no cost. This can be found at BLAST+ executables. There are also commercial programs available for purchase. Databases can be found from the NCBI site, as well as from Index of BLAST databases (FTP).

Process

Using a heuristic method, BLAST finds similar sequences, by locating short matches between the two sequences. This process of finding similar sequences is called seeding. It is after this first match that BLAST begins to make local alignments. While attempting to find similarity in sequences, sets of common letters, known as words, are very important. For example, suppose that the sequence contains the following stretch of letters, GLKFA. If a BLAST was being conducted under normal conditions, the word size would be 3 letters. In this case, using the given stretch of letters, the searched words would be GLK, LKF, KFA. The heuristic algorithm of BLAST locates all common three-letter words between the sequence of interest and the hit sequence or sequences from the database. This result will then be used to build an alignment. After making words for the sequence of interest, the rest of the words are also assembled. These words must satisfy a requirement of having a score of at least the threshold *T*, when compared by using a scoring matrix. One commonly used scoring matrix for BLAST searches is BLOSUM62, although the optimal scoring matrix depends on sequence similarity. Once both words and neighborhood words are assembled and compiled, they are compared to the sequences in the database in order to find matches. The threshold score *T* determines whether or not a particular word will be included in the alignment. Once seeding has been conducted, the alignment which is only 3 residues long, is extended in both directions by the algorithm used by BLAST. Each extension impacts the score of the alignment by either increasing or decreasing it. If this score is higher than a pre-determined *T*, the alignment will be included in the results given by BLAST. However, if this score is lower than this pre-determined *T*, the alignment will cease to extend, preventing the areas of poor alignment from being included in the BLAST results. Note that increasing the *T* score limits the amount of space available to search, decreasing the number of neighborhood words, while at the same time speeding up the process of BLAST.

Algorithm

To run the software, BLAST requires a query sequence to search for, and a sequence to search against (also called the target sequence) or a sequence database containing multiple such sequences. BLAST will find sub-sequences in the database which are similar to sub sequences in the query. In typical usage, the query sequence is much smaller than the database, e.g., the query may be one thousand nucleotides while the database is several billion nucleotides.

The main idea of BLAST is that there are often High-scoring Segment Pairs (HSP) contained in a statistically significant alignment. BLAST searches for high scoring sequence alignments between the query sequence and the existing sequences in the database using a heuristic approach that approximates the Smith-Waterman algorithm. However, the exhaustive Smith-Waterman approach is too slow for searching large genomic databases such as GenBank. Therefore, the BLAST algorithm uses a heuristic approach that is less accurate than the Smith-Waterman algorithm but over 50 times faster. The speed and relatively good accuracy of BLAST are among the key technical innovations of the BLAST programs.

An overview of the BLAST algorithm (a protein to protein search) is as follows: and CTGA2016.

1. Remove low-complexity region or sequence repeats in the query sequence.

"Low-complexity region" means a region of a sequence composed of few kinds of elements. These regions might give high scores that confuse the program to find the actual significant sequences in the database, so they should be filtered out. The regions will be marked with an X (protein sequences) or N (nucleic acid sequences) and then be ignored by the BLAST program. To filter out the low-complexity regions, the SEG program is used for protein sequences and the program DUST is used for DNA sequences. On the other hand, the program XNU is used to mask off the tandem repeats in protein sequences.

2. Make a *k*-letter word list of the query sequence.

Take k=3 for example, we list the words of length 3 in the query protein sequence (k is usually 11 for a DNA sequence) "sequentially", until the last letter of the query sequence is included. The method is illustrated in figure 1.

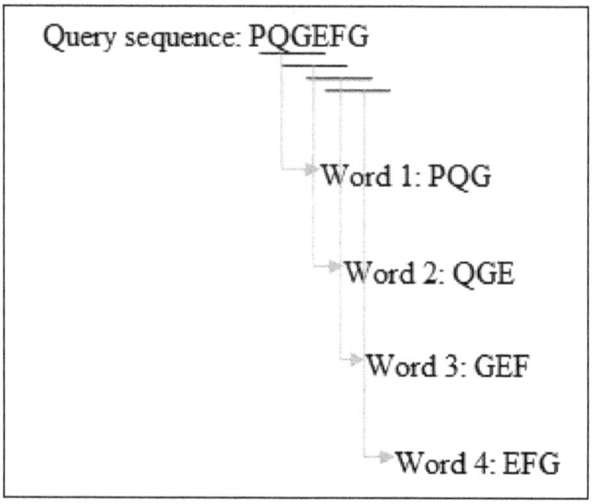

The method to establish the *k*-letter query word list. Adapted from Biological Sequence Analysis I, Current Topics in Genome Analysis .

3. List the possible matching words.

This step is one of the main differences between BLAST and FASTA. FASTA cares about all of the common words in the database and query sequences that are listed in step 2; however, BLAST only cares about the high-scoring words. The scores are created by comparing the word in the list in step 2 with all the 3-letter words. By using the scoring matrix (substitution matrix) to score the comparison of each residue pair, there are 20^3 possible match scores for a 3-letter word. For example, the score obtained by comparing PQG with PEG and PQA is 15 and 12, respectively. For DNA words, a match is scored as +5 and a mismatch as -4, or as +2 and -3. After that, a neighborhood word score threshold *T* is used to reduce the number of possible matching words. The words whose scores are greater than the threshold *T* will remain in the possible matching words list, while those with lower scores will be discarded. For example, PEG is kept, but PQA is abandoned when T is 13.

4. Organize the remaining high-scoring words into an efficient search tree.

This allows the program to rapidly compare the high-scoring words to the database sequences.

5. Repeat step 3 to 4 for each *k*-letter word in the query sequence.
6. Scan the database sequences for exact matches with the remaining high-scoring words.

The BLAST program scans the database sequences for the remaining high-scoring word, such as PEG, of each position. If an exact match is found, this match is used to seed a possible un-gapped alignment between the query and database sequences.

7. Extend the exact matches to high-scoring segment pair (HSP):

   - The original version of BLAST stretches a longer alignment between the query and the database sequence in the left and right directions, from the position where the exact match occurred. The extension does not stop until the accumulated total score of the HSP begins to decrease. A simplified example is presented in figure 2.

Query sequence: R P P Q G L F

Database sequence: D P **P E G** V V

→Exact match is scanned.

Score: -2 **7 7 2 6 1** -1

→HSP

Optimal accumulated score = 7+7+2+6+1 = 23

Fig. 2 The process to extend the exact match. Adapted from Biological Sequence Analysis I, Current Topics in Genome Analysis.

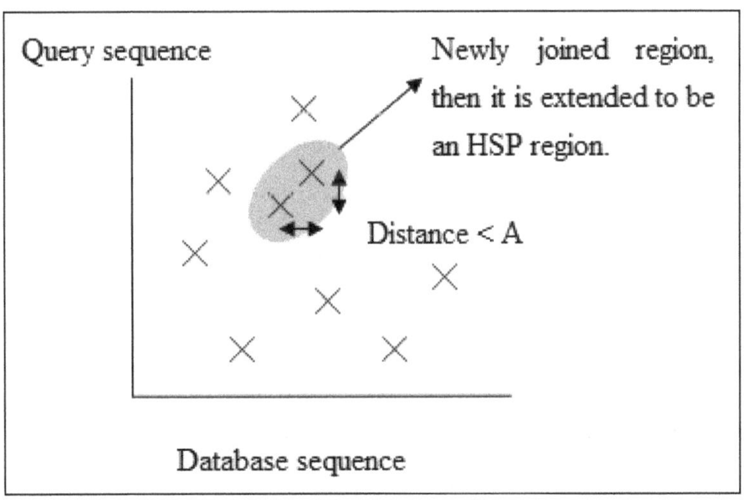

The positions of the exact matches.

   - To save more time, a newer version of BLAST, called BLAST2 or gapped BLAST, has been developed. BLAST2 adopts a lower neighborhood word score threshold to maintain the same level of sensitivity for detecting sequence similarity. Therefore, the possible matching words list in step 3 becomes longer. Next, the exact matched regions, within distance A from each other on the same diagonal in figure 3, will be joined as a longer new region. Finally, the new regions are then extended by the same method as in the original version of BLAST, and the HSPs' (High-scoring segment pair) scores of the extended regions are then created by using a substitution matrix as before.

8. List all of the HSPs in the database whose score is high enough to be considered.

We list the HSPs whose scores are greater than the empirically determined cutoff score *S*. By examining the distribution of the alignment scores modeled by comparing random sequences, a cutoff score *S* can be determined such that its value is large enough to guarantee the significance of the remaining HSPs.

9. Evaluate the significance of the HSP score.

BLAST next assesses the statistical significance of each HSP score by exploiting the Gumbel extreme value distribution (EVD). (It is proved that the distribution of Smith-Waterman local alignment scores between two random sequences follows the Gumbel EVD. For local alignments containing gaps it is not proved.). In accordance with the Gumbel EVD, the probability *p* of observing a score *S* equal to or greater than x is given by the equation,

$$p(S \geq x) = 1 - \exp\left(-e^{-\lambda(x-\mu)}\right)$$

where

$$\mu = {}^{\left[\log(Km'n')\right]}/{}_{\lambda}$$

The statistical parameters λ and K are estimated by fitting the distribution of the un-gapped local alignment scores, of the query sequence and a lot of shuffled versions (Global or local shuffling) of a database sequence, to the Gumbel extreme value distribution. Note that λ and K depend upon the substitution matrix, gap penalties, and sequence composition (the letter frequencies). m' and n' are the effective lengths of the query and database sequences, respectively. The original sequence length is shortened to the effective length to compensate for the edge effect (an alignment start near the end of one of the query or database sequence is likely not to have enough sequence to build an optimal alignment). They can be calculated as,

$$m' \approx m - {}^{(\ln Kmn)}/{}_{H}$$

$$n' \approx n - {}^{(\ln Kmn)}/{}_{H}$$

where H is the average expected score per aligned pair of residues in an alignment of two random sequences. Altschul and Gish gave the typical values, $\lambda = 0.318,\ \mathrm{K} = 0.13$, and $\mathrm{H} = 0.40$, for un-gapped local alignment using BLOSUM62 as the substitution matrix. Using the typical values for assessing the significance is called the lookup table method; it is not accurate. The expect score *E* of a database match is the number of times that an unrelated database sequence would obtain a score S higher than x by chance. The expectation E obtained in a search for a database of D sequences is given by,

$$E \approx 1 - e^{-p(s>x)D}$$

Furthermore, when $p < 0.1$, E could be approximated by the Poisson distribution as,

$$E \approx pD$$

This expectation or expect value "E" (often called an E score or E-value or e-value) assessing the significance of the HSP score for un-gapped local alignment is reported in the BLAST results. The calculation shown here is modified if individual HSPs are combined, such as when producing gapped alignments (described below), due to the variation of the statistical parameters.

10. Make two or more HSP regions into a longer alignment.

Sometimes, we find two or more HSP regions in one database sequence that can be made into a longer alignment. This provides additional evidence of the relation between the query and database sequence. There are two methods, the Poisson method and the sum-of-scores method, to compare the significance of the newly combined HSP regions. Suppose that there are two combined HSP regions with the pairs of scores (65, 40) and (52, 45), respectively. The Poisson method gives more significance to the set with the maximal lower score (45>40). However, the sum-of-scores method prefers the first set, because 65+40 (105) is greater than 52+45(97). The original BLAST uses the Poisson method; gapped BLAST and the WU-BLAST uses the sum-of scores method.

11. Show the gapped Smith-Waterman local alignments of the query and each of the matched database sequences.
    - The original BLAST only generates un-gapped alignments including the initially found HSPs individually, even when there is more than one HSP found in one database sequence.
    - BLAST2 produces a single alignment with gaps that can include all of the initially found HSP regions. Note that the computation of the score and its corresponding E-value involves use of adequate gap penalties.
12. Report every match whose expect score is lower than a threshold parameter ***E***.

Parallel BLAST

Parallel BLAST versions of split databases are implemented using MPI and Pthreads, and have been ported to various platforms including Windows, Linux, Solaris, Mac OS X, and AIX. Popular approaches to parallelize BLAST include query distribution, hash table segmentation, computation parallelization, and database segmentation (partition). Databases are split into equal sized pieces and stored locally on each node. Each query is run on all nodes in parallel and the resultant BLAST output files from all nodes merged to yield the final output.

Program

The BLAST program can either be downloaded and run as a command-line utility "blastall" or accessed for free over the web. The BLAST web server, hosted by the NCBI, allows anyone with a web browser to perform similarity searches against constantly updated databases of proteins and DNA that include most of the newly sequenced organisms.

The BLAST program is based on an open-source format, giving everyone access to it and enabling them to have the ability to change the program code. This has led to the creation of several BLAST "spin-offs".

There are now a handful of different BLAST programs available, which can be used depending on what one is attempting to do and what they are working with. These different programs vary in query sequence input, the database being searched, and what is being compared. These programs and their details are listed below:

BLAST is actually a family of programs (all included in the blastall executable). These include:

Nucleotide-nucleotide BLAST (blastn):

> This program, given a DNA query, returns the most similar DNA sequences from the DNA database that the user specifies.

Protein-protein BLAST (blastp):

> This program, given a protein query, returns the most similar protein sequences from the protein database that the user specifies.

Position-Specific Iterative BLAST (PSI-BLAST) (blastpgp):

> This program is used to find distant relatives of a protein. First, a list of all closely related proteins is created. These proteins are combined into a general "profile" sequence, which summarises significant features present in these sequences. A query against the protein database is then run using this profile, and a larger group of proteins is found. This larger group is used to construct another profile, and the process is repeated.
>
> By including related proteins in the search, PSI-BLAST is much more sensitive in picking up distant evolutionary relationships than a standard protein-protein BLAST.

Nucleotide 6-frame translation-protein (blastx):

> This program compares the six-frame conceptual translation products of a nucleotide query sequence (both strands) against a protein sequence database.

Nucleotide 6-frame translation-nucleotide 6-frame translation (tblastx):

> This program is the slowest of the BLAST family. It translates the query nucleotide sequence in all six possible frames and compares it against the six-frame translations of a nucleotide sequence database. The purpose of tblastx is to find very distant relationships between nucleotide sequences.

Protein-nucleotide 6-frame translation (tblastn):

> This program compares a protein query against the all six reading frames of a nucleotide sequence database.

Large numbers of query sequences (megablast):

> When comparing large numbers of input sequences via the command-line BLAST, "megablast" is much faster than running BLAST multiple times. It concatenates many input sequences together to form a large sequence before searching the BLAST database, then post-analyzes the search results to glean individual alignments and statistical values.

Of these programs, BLASTn and BLASTp are the most commonly used because they use direct comparisons, and do not require translations. However, since protein sequences are better conserved evolutionarily than nucleotide sequences, tBLASTn, tBLASTx, and BLASTx, produce more reliable and accurate results when dealing with coding DNA. They also enable one to be able to directly see the function of the protein sequence, since by translating the sequence of interest before searching often gives you annotated protein hits.

Alternative Versions

A version designed for comparing large genomes or DNA is BLASTZ.

CS-BLAST (ContSxt-Specific BLAST) is an extended version of BLAST for searching protein sequences that finds twice as many remotely related sequences as BLAST at the same speed and error rate. In CS-BLAST, the mutation probabilities between amino acids depend not only on the single amino acid, as in BLAST, but also on its local sequence context. Washington University produced an alternative version of NCBI BLAST, called WU-BLAST. The rights have since been acquired to Advanced Biocomputing, LLC.

In 2009, NCBI has released a new set of BLAST executables, the C++ based BLAST+, and has released parallel versions until 2.2.26. Starting with version 2.2.27 (April

2013), only BLAST+ executables are available. Among the changes is the replacement of the blastall executable with separate executables for the different BLAST programs, and changes in option handling. The formatdb utility (C based) has been replaced by makeblastdb (C++ based) and databases formatted by either one should be compatible for identical blast releases. The algorithms remain similar, however, the number of hits found and their order can vary significantly between the older and the newer version.

Accelerated Versions

- CLC bio and SciEngines GmbH collaborate on an FPGA accelerator they claim will give 188x acceleration of BLAST.
- TimeLogic offers another FPGA-accelerated implementation of the BLAST algorithm called Tera-BLAST.
- The Mitrion-C Open Bio Project is an ongoing effort to port BLAST to run on Mitrion FPGAs.
- The GPU-Blast is an accelerated version of NCBI BLASTP for CUDA which is 3x~4x faster than NCBI Blast.
- The CUDA-BLASTP is a version of BLASTP that is GPU-accelerated and is claimed to run up to 10x faster than NCBI BLAST.
- G-BLASTN is an accelerated version of NCBI blastn and megablast, whose speedup varies from 4x to 14x (compared to the same runs with 4 CPU threads). Its current limitation is that the database must fit into the GPU memory.
- MPIBlast is a parallel implementation of NCBI BLAST using Message Passing Interface. By efficiently utilizing distributed computational resources through database fragmentation, query segmentation, intelligent scheduling, and parallel I/O, mpiBLAST improves NCBI BLAST performance by several orders of magnitude while scaling to hundreds of processors.
- Paracel BLAST is a commercial parallel implementation of NCBI BLAST, supporting hundreds of processors.
- CaBLAST makes search on large databases orders of magnitude faster by exploiting redundancy in data.

Alternatives to BLAST

An extremely fast but considerably less sensitive alternative to BLAST is BLAT (Blast Like Alignment Tool). While BLAST does a linear search, BLAT relies on k-mer indexing the database, and can thus often find seeds faster. Another software alternative similar to BLAT is PatternHunter.

Advances in sequencing technology in the late 2000s has made searching for very similar nucleotide matches an important problem. New alignment programs tailored for this use typically use BWT-indexing of the target database (typically a genome). Input sequences can then be mapped very quickly, and output is typically in the form of a BAM file. Example alignment programs are BWA, SOAP, and Bowtie.

For protein identification, searching for known domains (for instance from Pfam) by matching with Hidden Markov Models is a popular alternative, such as HMMER.

An alternative to BLAST for comparing two banks of sequences is PLAST. PLAST provides a high-performance general purpose bank to bank sequence similarity search tool relying on the PLAST and ORIS algorithms. Results of PLAST are very similar to BLAST, but KLAST is significantly faster and capable of comparing large sets of sequences with a small memory (i.e. RAM) footprint.

For applications in metagenomics, where the task is to compare billions of short DNA reads against tens of millions of protein references, DIAMOND runs at up to 20,000 times as fast as BLASTX, while maintaining a high level of sensitivity.

Uses of BLAST

BLAST can be used for several purposes. These include identifying species, locating domains, establishing phylogeny, DNA mapping, and comparison.

Identifying species:

> With the use of BLAST, you can possibly correctly identify a species or find homologous species. This can be useful, for example, when you are working with a DNA sequence from an unknown species.

Locating domains:

> When working with a protein sequence you can input it into BLAST, to locate known domains within the sequence of interest.

Establishing phylogeny:

> Using the results received through BLAST you can create a phylogenetic tree using the BLAST web-page. Phylogenies based on BLAST alone are less reliable than other purpose-built computational phylogenetic methods, so should only be relied upon for "first pass" phylogenetic analyses.

DNA mapping:

> When working with a known species, and looking to sequence a gene at an unknown location, BLAST can compare the chromosomal position of the sequence of interest, to relevant sequences in the database(s).

Comparison:

> When working with genes, BLAST can locate common genes in two related species, and can be used to map annotations from one organism to another.

Comparing BLAST and the Smith-Waterman Process

While both Smith-Waterman and BLAST are used to find homologous sequences by searching and comparing a query sequence with those in the databases, they do have their differences.

Due to the fact that BLAST is based on a heuristic algorithm, the results received through BLAST, in terms of the hits found, may not be the best possible results, as it will not provide you with all the hits within the database. BLAST misses hard to find matches.

A better alternative in order to find the best possible results would be to use the Smith-Waterman algorithm. This method varies from the BLAST method in two areas, accuracy and speed. The Smith-Waterman option provides better accuracy, in that it finds matches that BLAST cannot, because it does not miss any information. Therefore, it is necessary for remote homology. However, when compared to BLAST, it is more time consuming, not to mention that it requires large amounts of computer usage and space. However, technologies to speed up the Smith-Waterman process have been found to improve the time necessary to perform a search dramatically. These technologies include FPGA chips and SIMD technology.

In order to receive better results from BLAST, the settings can be changed from their default settings. However, there is no given or set way of changing these settings in order to receive the best results for a given sequence. The settings available for change are E-Value, gap costs, filters, word size, and substitution matrix. Note, that the algorithm used for BLAST was developed from the algorithm used for Smith-Waterman. BLAST employs an alignment which finds "local alignments between sequences by finding short matches and from these initial matches (local) alignments are created".

Multiple Sequence Alignment

A multiple sequence alignment (MSA) is a sequence alignment of three or more biological sequences, generally protein, DNA, or RNA. In many cases, the input set of query sequences are assumed to have an evolutionary relationship by which they share a lineage and are descended from a common ancestor. From the resulting MSA, sequence homology can be inferred and phylogenetic analysis can be conducted to assess the sequences' shared evolutionary origins. Visual depictions of the alignment as in the image at right illustrate mutation events such as point mutations (single amino acid or nucleotide changes) that appear as differing characters in a single alignment column, and insertion or deletion mutations (indels or gaps) that appear as hyphens in one or more of the sequences in the alignment. Multiple sequence alignment is often used to assess sequence conservation of protein domains, tertiary and secondary structures, and even individual amino acids or nucleotides.

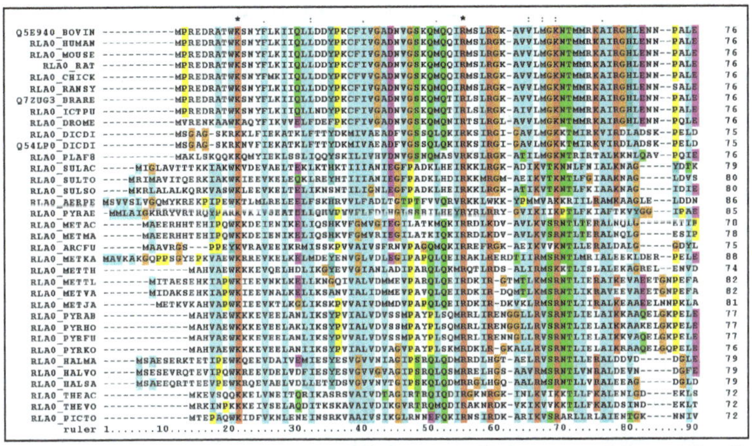

First 90 positions of a protein multiple sequence alignment of instances of the acidic ribosomal protein P0 (L10E) from several organisms. Generated with ClustalX.

Multiple sequence alignment also refers to the process of aligning such a sequence set. Because three or more sequences of biologically relevant length can be difficult and are almost always time-consuming to align by hand, computational algorithms are used to produce and analyze the alignments. MSAs require more sophisticated methodologies than pairwise alignment because they are more computationally complex. Most multiple sequence alignment programs use heuristic methods rather than global optimization because identifying the optimal alignment between more than a few sequences of moderate length is prohibitively computationally expensive.

Dynamic Programming and Computational Complexity

A direct method for producing an MSA uses the dynamic programming technique to identify the globally optimal alignment solution. For proteins, this method usually involves two sets of parameters: a gap penalty and a substitution matrix assigning scores or probabilities to the alignment of each possible pair of amino acids based on the similarity of the amino acids' chemical properties and the evolutionary probability of the mutation. For nucleotide sequences a similar gap penalty is used, but a much simpler substitution matrix, wherein only identical matches and mismatches are considered, is typical. The scores in the substitution matrix may be either all positive or a mix of positive and negative in the case of a global alignment, but must be both positive and negative, in the case of a local alignment.

For n individual sequences, the naive method requires constructing the n-dimensional equivalent of the matrix formed in standard pairwise sequence alignment. The search space thus increases exponentially with increasing *n* and is also strongly dependent on sequence length. Expressed with the big O notation commonly used to measure computational complexity, a naïve MSA takes *O(Length)* time to produce. To find the global optimum for *n* sequences this way has been shown to be an NP-complete problem. In 1989, based on Carrillo-Lipman Algorithm, Altschul introduced a practical method that uses pairwise alignments to constrain the n-dimensional search space. In this approach pairwise dynamic programming alignments are performed on each pair of sequences in the query set, and only the space near the n-dimensional intersection of these alignments is searched for the n-way alignment. The MSA program optimizes the sum of all of the pairs of characters at each position in the alignment (the so-called *sum of pair* score) and has been implemented in a software program for constructing multiple sequence alignments.

Progressive Alignment Construction

The most widely used approach to multiple sequence alignments uses a heuristic search known as progressive technique (also known as the hierarchical or tree method) developed by Paulien Hogeweg and Ben Hesper in 1984. Progressive alignment builds up a final MSA by combining pairwise alignments beginning with the most similar pair and progressing to the most distantly related. All progressive alignment methods require two stages: a first stage in which the relationships between the sequences are represented as a tree, called a *guide tree*, and a second step in which the MSA is built by adding the sequences sequentially to the growing MSA according to the guide tree. The initial *guide tree* is determined by an efficient clustering method such as neighbor-joining or UPGMA, and may use distances based on the number of identical two letter sub-sequences (as in FASTA rather than a dynamic programming alignment).

Progressive alignments are not guaranteed to be globally optimal. The primary problem is that when errors are made at any stage in growing the MSA, these errors are then propagated through to the final result. Performance is also particularly bad when all of the sequences in the set are rather distantly related. Most modern progressive methods modify their scoring function with a secondary weighting function that assigns scaling factors to individual members of the query set in a nonlinear fashion based on their phylogenetic distance from their nearest neighbors. This corrects for non-random selection of the sequences given to the alignment program.

Progressive alignment methods are efficient enough to implement on a large scale for many (100s to 1000s) sequences. Progressive alignment services are commonly available on publicly accessible web servers so users need not locally install the applications of interest. The most popular progressive alignment method has been the Clustal family, especially the weighted variant ClustalW to which access is provided by a large number of web portals including GenomeNet, EBI, and EMBNet. Different portals or implementations can vary in user interface and make different parameters accessible to the user. ClustalW is used extensively for phylogenetic tree construction, in spite of the author's explicit warnings that unedited alignments should not be used in such studies and as input for protein structure prediction by homology modeling. Current version of Clustal family is ClustalW2. EMBL-EBI announced that CLustalW2 will be expired in August 2015. They recommend Clustal Omega which performs based on seeded guide trees and HMM profile-profile techniques for protein alignments. They offer different MSA tools for progressive DNA alignments. One of them is MAFFT (Multiple Alignment using Fast Fourier Transform).

Another common progressive alignment method called T-Coffee is slower than Clustal and its derivatives but generally produces more accurate alignments for distantly related sequence sets. T-Coffee calculates pairwise alignments by combining the direct alignment of the pair with indirect alignments that aligns each sequence of the pair to a third sequence. It uses the output from Clustal as well as another local alignment program LALIGN, which finds multiple regions of local alignment between two sequences. The resulting alignment and phylogenetic tree are used as a guide to produce new and more accurate weighting factors.

Because progressive methods are heuristics that are not guaranteed to converge to a global optimum, alignment quality can be difficult to evaluate and their true biological significance can be obscure. A semi-progressive method that improves alignment quality and does not use a lossy heuristic while still running in polynomial time has been implemented in the program PSAlign.

Iterative Methods

A set of methods to produce MSAs while reducing the errors inherent in progressive methods are classified as "iterative" because they work similarly to progressive methods but repeatedly realign the initial sequences as well as adding new sequences to the growing MSA. One reason progressive methods are so strongly dependent on a high-quality initial alignment is the fact that these alignments are always incorporated into the final result — that is, once a sequence has been aligned into the MSA, its alignment is not considered further. This approximation improves efficiency at the cost of accuracy. By contrast, iterative methods can return to previously calculated pairwise alignments or sub-MSAs incorporating subsets of the query sequence as a means of optimizing a general objective function such as finding a high-quality alignment score.

A variety of subtly different iteration methods have been implemented and made available in software packages; reviews and comparisons have been useful but generally refrain from choosing a "best" technique. The software package PRRN/PRRP uses a hill-climbing algorithm to optimize its MSA alignment score and iteratively corrects both alignment weights and locally divergent or "gappy" regions of the growing MSA. PRRP performs best when refining an alignment previously constructed by a faster method.

Another iterative program, DIALIGN, takes an unusual approach of focusing narrowly on local alignments between sub-segments or sequence motifs without introducing a gap penalty. The alignment of individual motifs is then achieved with a matrix representation similar to a dot-matrix plot in a pairwise alignment. An alternative method that uses fast local alignments as anchor points or "seeds" for a slower global-alignment procedure is implemented in the CHAOS/DIALIGN suite.

A third popular iteration-based method called MUSCLE (multiple sequence alignment by log-expectation) improves on progressive methods with a more accurate distance measure to assess the relatedness of two sequences. The distance measure is updated between iteration stages (although, in its original form, MUSCLE contained only 2-3 iterations depending on whether refinement was enabled).

Consensus Methods

Consensus methods attempt to find the optimal multiple sequence alignment given multiple different alignments of the same set of sequences. There are two commonly used consensus methods, M-COFFEE and MergeAlign. M-COFFEE uses multiple sequence alignments generated by seven different methods to generate consensus alignments. MergeAlign is capable of generating consensus alignments from any number of input alignments generated using different models of sequence evolution or different methods of multiple sequence alignment. The default option for MergeAlign is to infer a consensus alignment using alignments generated using 91 different models of protein sequence evolution.

Hidden Markov Models

Hidden Markov models are probabilistic models that can assign likelihoods to all possible combinations of gaps, matches, and mismatches to determine the most likely MSA or set of possible MSAs. HMMs can produce a single highest-scoring output but can also generate a family of possible alignments that can then be evaluated for biological significance. HMMs can produce both global and local alignments. Although HMM-based methods have been developed relatively recently, they offer significant improvements in computational speed, especially for sequences that contain overlapping regions.

Typical HMM-based methods work by representing an MSA as a form of directed acyclic graph known as a partial-order graph, which consists of a series of nodes representing possible entries in the columns of an MSA. In this representation a column that is absolutely conserved (that is, that all the sequences in the MSA share a particular character at a particular position) is coded as a single node with as many outgoing connections as there are possible characters in the next column of the alignment. In the terms of a typical hidden Markov model, the observed states are the individual alignment columns and the "hidden" states represent the presumed ancestral sequence from which the sequences in the query set are hypothesized to have descended. An efficient search variant of the dynamic programming method, known as the Viterbi algorithm, is generally used to successively align the growing MSA to the next sequence in the query set to produce a new MSA. This is distinct from progressive alignment methods because the alignment of prior sequences is updated at each new sequence addition. However, like progressive methods, this technique can be influenced by the order in which the sequences in the query set are integrated into the alignment, especially when the sequences are distantly related.

Several software programs are available in which variants of HMM-based methods have been implemented and which are noted for their scalability and efficiency, although properly using an HMM method is more complex than using more common progressive methods. The simplest is POA (Partial-Order Alignment); a similar but more generalized method is implemented in the packages SAM (Sequence Alignment and Modeling System). and HMMER. SAM has been used as a source of alignments for protein structure prediction to participate in the CASP structure prediction experiment and to develop a database of predicted proteins in the yeast species *S. cerevisiae*. HHsearch is a software package for the detection of remotely related protein sequences based on the pairwise comparison of HMMs. A server running HHsearch (HHpred) was by far the fastest of the 10 best automatic structure prediction servers in the CASP7 and CASP8 structure prediction competitions.

Genetic Algorithms and Simulated Annealing

Standard optimization techniques in computer science — both of which were inspired by, but do not directly reproduce, physical processes — have also been used in an attempt to more efficiently produce quality MSAs. One such technique, genetic algorithms, has been used for MSA production in an attempt to broadly simulate the hypothesized evolutionary process that gave rise to the divergence in the query set. The method works by breaking a series of possible MSAs into fragments and repeatedly rearranging those fragments with the introduction of gaps at varying positions. A general objective function is optimized during the simulation, most generally the "sum of pairs" maximization function introduced in dynamic programming-based MSA methods. A technique for protein sequences has been implemented in the software program SAGA (Sequence Alignment by Genetic Algorithm) and its equivalent in RNA is called RAGA.

The technique of simulated annealing, by which an existing MSA produced by another method is refined by a series of rearrangements designed to find better regions of alignment space than the one the input alignment already occupies. Like the genetic algorithm method, simulated annealing maximizes an objective function like the sum-of-pairs function. Simulated annealing uses a metaphorical "temperature factor" that determines the rate at which rearrangements proceed and the likelihood of each rearrangement; typical usage alternates periods of high rearrangement rates with relatively low likelihood (to explore more distant regions of alignment space) with periods of lower rates and higher likelihoods to more thoroughly explore local minima near the newly "colonized" regions. This approach has been implemented in the program MSA-SA (Multiple Sequence Alignment by Simulated Annealing).

Phylogeny-aware Methods

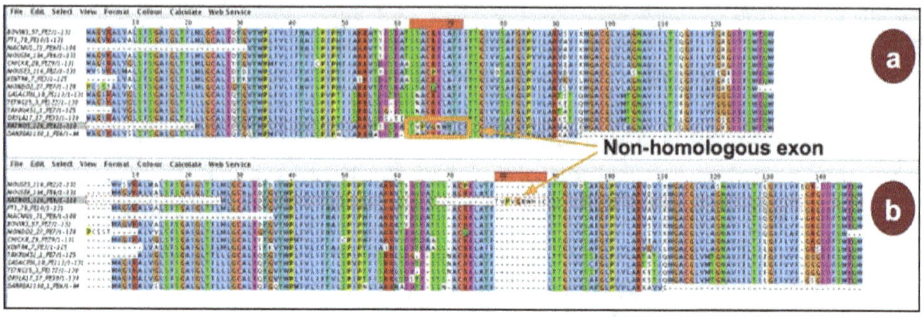

Non-homologous exon alignment by an iterative method (a), and by a phylogeny-aware method (b).

Most multiple sequence alignment methods try to minimize the number of insertions/ deletions (gaps) and, as a consequence, produce compact alignments. This causes several problems if the sequences to be aligned contain non-homologous regions, if gaps are informative in a phylogeny analysis. These problems are common in newly produced sequences that are poorly annotated and may contain frame-shifts, wrong domains or non-homologous spliced exons.

The first such method was developed in 2005 by Löytynoja and Goldman. The same authors released a software package called *PRANK* in 2008. PRANK improves alignments when insertions are present. Nevertheless, it runs slowly compared to progressive and/ or iterative methods which have been developed for several years.

In 2012, two new phylogeny-aware tools appeared. One is called *PAGAN* that was developed by the same team as PRANK. The other is *ProGraphMSA* developed by Szalkowski. Both software packages were developed independently but share common features, notably the use of graph algorithms to improve the recognition of non-homologous regions, and an improvement in code making these software faster than PRANK.

Motif Finding

```
DAMMfly2R_ : MYLPERTEHQKIERLY-----------------------------------------------DSNRVN--------------AEPGQGL----
DCP1fly2R_ : ----------------MTD-----------------ECVTRNYGVGIRSPNGSENRGS-FIMADNTDAK--------------GCTPESLVVGG
DRICEfly3R : MDATNNGESADQVGIRVGN-----------------PEQPNDHTDALGSV-GSGGAGSSGLVAGSSHPY--------------GSGAIGQLANG
DECAYfly3R : MDDTDFSLFGQKNKHK-------------------------------------------KDKADATKIA--------------HTPTSEL----
DRONCfly3L : MQPPELEIGMPKRHREHIRKNLNILVEWTNYERLAMECVQQGILTVQMLRNTQDLNGK-PFNMDEKDVRVEQHRRLLLKITQRGPTAYNLLINA
STRICAfly2 : MGWWSKKSETDRSQPSQELVAQDPRTRVQTTSAATETTNTAVQNSTITDNNKQTVTFI-TTRQTVTHTQRALITETTTRRTPSQAELEALFAKI
DREDDPAfly : MSASAIYRPFPKVKHFCIFPIAMAGSNLLIHLDTIDQNDLIYVERDMNFAQKVGLCFL-LYGDDHSDATYILQKLLAMTRSDFPQSDLLIKFAK
DREDDPBfly : MSASAIYRPFPKVKHFCIFPIAMAGSNLLIHLDTIDQNDLIYVERDMNFAQKVGLCFL-LYGDDHSDATYILQKLLAMTRSDFPQSDLLIKFAK
DREDDPCfly : MSASAIYRPFPKVKHFCIFPIAMAGSNLLIHLDTIDQNDLIYVERDMNFAQKVGLCFL-LYGDDHSDATYILQKLLAMTRSDFPQSDLLIKFAK
```

Alignment of the seven Drosophila caspases colored by motifs as identified by MEME. When motif positions and sequence alignments are generated independently, they often correlate well but not perfectly, as in this example.

Motif finding, also known as profile analysis, is a method of locating sequence motifs in global MSAs that is both a means of producing a better MSA and a means of producing a scoring matrix for use in searching other sequences for similar motifs. A variety of methods for isolating the motifs have been developed, but all are based on identifying short highly conserved patterns within the larger alignment and constructing a matrix similar to a substitution matrix that reflects the amino acid or nucleotide composition of each position in the putative motif. The alignment can then be refined using these matrices. In standard profile analysis, the matrix includes entries for each possible character as well as entries for gaps. Alternatively, statistical pattern-finding algorithms can identify motifs as a precursor to an MSA rather than as a derivation. In many cases when the query set contains only a small number of sequences or contains only highly related sequences, pseudocounts are added to normalize the distribution reflected in the scoring matrix. In particular, this corrects zero-probability entries in the matrix to values that are small but nonzero.

Blocks analysis is a method of motif finding that restricts motifs to ungapped regions in the alignment. Blocks can be generated from an MSA or they can be extracted from unaligned sequences using a precalculated set of common motifs previously generated from known gene families. Block scoring generally relies on the spacing of high-frequency characters rather than on the calculation of an explicit substitution matrix. The BLOCKS server provides an interactive method to locate such motifs in unaligned sequences.

Statistical pattern-matching has been implemented using both the expectation-maximization algorithm and the Gibbs sampler. One of the most common motif-finding tools, known as MEME, uses expectation maximization and hidden Markov methods to generate motifs that are then used as search tools by its companion MAST in the combined suite MEME/MAST.

Non-coding Multiple Sequence Alignment

Non-coding DNA regions, especially TFBSs, are rather more conserved and not necessarily evolutionarily related, and may have converged from non-common ancestors. Thus, the assumptions used to align protein sequences and DNA coding regions are inherently different from those that hold for TFBS sequences. Although it is meaningful to align DNA coding regions for homologous sequences using mutation operators, alignment of binding site sequences for the same transcription factor cannot rely on evolutionary related mutation operations. Similarly, the evolutionary operator of point mutations can be used to define an edit distance for coding sequences, but this has little meaning for TFBS sequences because any sequence variation has to maintain a certain level of specificity for the binding site to function. This becomes specifically important when trying to align known TFBS sequences to build supervised models to predict unknown locations of the same TFBS. Hence, Multiple Sequence Alignment methods need to adjust the underlying evolutionary hypothesis and the operators used as in the work published incorporating neighbouring base thermodynamic information to align the binding sites searching for the lowest thermodynamic alignment conserving specificity of the binding site, EDNA.

Alignment Visualization and Quality Control

The necessary use of heuristics for multiple alignment means that for an arbitrary set of proteins, there is always a good chance that an alignment will contain errors. For example, an evaluation of several leading alignment programs using the BAliBase benchmark found that at least 24% of all pairs of aligned amino acids were incorrectly aligned. These errors can arise because of unique insertions into one or more regions of sequences, or through some more complex evolutionary process leading to proteins that do not align easily by sequence alone. As the number of sequence and their divergence increases many more errors will be made simply because of the heuristic nature of MSA algorithms. Multiple sequence alignment viewers enable alignments to be visually reviewed, often by inspecting the quality of alignment for annotated functional sites on two or more sequences. Many also enable the alignment to be edited to correct these (usually minor) errors, in order to obtain an optimal 'curated' alignment suitable for use in phylogenetic analysis or comparative modeling.

However, as the number of sequences increases and especially in genome-wide studies that involve many MSAs it is impossible to manually curate all alignments. Furthermore, manual curation is subjective. And finally, even the best expert cannot confidently align the more ambiguous cases of highly diverged sequences. In such cases it is common practice to use automatic procedures to exclude unreliably aligned regions from the MSA. For the purpose of phylogeny reconstruction the Gblocks program is widely used to remove alignment blocks suspect of low quality, according to various cutoffs on the number of gapped sequences in alignment columns. However, these criteria may excessively filter out regions with insertion/deletion events that may still be aligned reliably, and these regions might be desirable for other purposes such as detection of positive selection. A few alignment algorithms output site-specific scores that allow the selection of high-confidence regions. Such a service was first offered by the SOAP program, which tests the robustness of each column to perturbation in the parameters of the popular alignment program CLUSTALW. The T-Coffee program uses a library of alignments in the construction of the final MSA, and its output MSA is colored according to confidence scores that reflect the agreement between different alignments in the library regarding each aligned residue. Its extension, TCS : (Transitive Consistency Score), uses T-Coffee libraries of pairwise alignments to evaluate any third party MSA. Pairwise projections can be produced using fast or slow methods, thus allowing a trade-off between speed and accuracy. Another alignment program that can output an MSA with confidence scores is FSA, which uses a statistical model that allows calculation of the uncertainty in the alignment. The HoT (Heads-Or-Tails) score can be used as a measure of site-specific alignment uncertainty due to the existence of multiple co-optimal solutions. The GUIDANCE program calculates a similar site-specific confidence measure based on the robustness of the alignment to uncertainty in the guide tree that is used in progressive alignment programs. An alternative, more statistically justified approach to assess alignment uncertainty is the use of probabilistic evolutionary models for joint estimation of phylogeny and alignment. A Bayesian approach allows calculation of posterior probabilities of estimated phylogeny and alignment, which is a measure of the confidence in these estimates. In this case, a posterior probability can be calculated for each site in the alignment. Such an approach was implemented in the program BAli-Phy.

There free available programs for visualization of multiple sequence alignments: JalView, UGENE.

Use in Phylogenetics

Multiple sequence alignments can be used to create a phylogenetic tree. This is made possible by two reasons. The first is because functional domains that are known in annotated sequences can be used for alignment in non-annotated sequences. The other is that conserved regions known to be functionally important can be found. This makes it possible for multiple sequence alignments to be used to analyze and find evolutionary relationships through homology between sequences. Point mutations and insertion or deletion events (called indels) can be detected.

Multiple sequence alignments can also be used to identify functionally important sites, such as binding sites, active sites, or sites corresponding to other key functions, by locating conserved domains. When looking at multiple sequence alignments, it is useful to consider different aspects of the sequences when comparing sequences. These aspects include identity, similarity, and homology. Identity means that the sequences have identical residues at their respective positions. On the other hand, similarity has to do with the sequences being compared having similar residues quantitatively. For example, in terms of nucleotide sequences, pyrimidines are considered similar to each other, as are purines. Similarity ultimately leads to homology, in that the more similar sequences are, the closer they are to being homologous. This similarity in sequences can then go on to help find common ancestry.

References

- Mount DM. (2004). Bioinformatics: Sequence and Genome Analysis (2nd ed.). Cold Spring Harbor Laboratory Press: Cold Spring Harbor, NY. ISBN 0-87969-608-7
- Brohée, Sylvain; Barriot, Roland; Moreau, Yves. "Biological knowledge bases using Wikis: combining the flexibility of Wikis with the structure of databases". Bioinformatics. Oxford Journals. Retrieved 11, August 2020
- Wong, KC (2016). Computational Biology and Bioinformatics: Gene Regulation. CRC Press (Taylor & Francis Group). ISBN 9781498724975
- Kim N; Lee C (2008). "Bioinformatics detection of alternative splicing". Methods Mol. Biol. Methods in Molecular Biology™. 452: 179–97. doi:10.1007/978-1-60327-159-2_9. ISBN 978-1-58829-707-5. PMID 18566765

Permissions

All chapters in this book are published with permission under the Creative Commons Attribution Share Alike License or equivalent. Every chapter published in this book has been scrutinized by our experts. Their significance has been extensively debated. The topics covered herein carry significant information for a comprehensive understanding. They may even be implemented as practical applications or may be referred to as a beginning point for further studies.

We would like to thank the editorial team for lending their expertise to make the book truly unique. They have played a crucial role in the development of this book. Without their invaluable contributions this book wouldn't have been possible. They have made vital efforts to compile up to date information on the varied aspects of this subject to make this book a valuable addition to the collection of many professionals and students.

This book was conceptualized with the vision of imparting up-to-date and integrated information in this field. To ensure the same, a matchless editorial board was set up. Every individual on the board went through rigorous rounds of assessment to prove their worth. After which they invested a large part of their time researching and compiling the most relevant data for our readers.

The editorial board has been involved in producing this book since its inception. They have spent rigorous hours researching and exploring the diverse topics which have resulted in the successful publishing of this book. They have passed on their knowledge of decades through this book. To expedite this challenging task, the publisher supported the team at every step. A small team of assistant editors was also appointed to further simplify the editing procedure and attain best results for the readers.

Apart from the editorial board, the designing team has also invested a significant amount of their time in understanding the subject and creating the most relevant covers. They scrutinized every image to scout for the most suitable representation of the subject and create an appropriate cover for the book.

The publishing team has been an ardent support to the editorial, designing and production team. Their endless efforts to recruit the best for this project, has resulted in the accomplishment of this book. They are a veteran in the field of academics and their pool of knowledge is as vast as their experience in printing. Their expertise and guidance has proved useful at every step. Their uncompromising quality standards have made this book an exceptional effort. Their encouragement from time to time has been an inspiration for everyone.

The publisher and the editorial board hope that this book will prove to be a valuable piece of knowledge for students, practitioners and scholars across the globe.

Index

www.ingramcontent.com/pod-product-compliance
Lightning Source LLC
LaVergne TN
LVHW080804220926
848485LV00011B/161